Gödel's Theorem: A Very Short Introduction

VERY SHORT INTRODUCTIONS are for anyone wanting a stimulating and accessible way into a new subject. They are written by experts, and have been translated into more than 45 different languages.

The series began in 1995, and now covers a wide variety of topics in every discipline. The VSI library currently contains over 700 volumes—a Very Short Introduction to everything from Psychology and Philosophy of Science to American History and Relativity—and continues to grow in every subject area.

Very Short Introductions available now:

ABOLITIONISM Richard S. Newman
THE ABRAHAMIC RELIGIONS
    Charles L. Cohen
ACCOUNTING Christopher Nobes
ADOLESCENCE Peter K. Smith
THEODOR W. ADORNO
    Andrew Bowie
ADVERTISING Winston Fletcher
AERIAL WARFARE Frank Ledwidge
AESTHETICS Bence Nanay
AFRICAN AMERICAN RELIGION
    Eddie S. Glaude Jr
AFRICAN HISTORY John Parker and
    Richard Rathbone
AFRICAN POLITICS Ian Taylor
AFRICAN RELIGIONS
    Jacob K. Olupona
AGEING Nancy A. Pachana
AGNOSTICISM Robin Le Poidevin
AGRICULTURE Paul Brassley and
    Richard Soffe
ALEXANDER THE GREAT
    Hugh Bowden
ALGEBRA Peter M. Higgins
AMERICAN BUSINESS HISTORY
    Walter A. Friedman
AMERICAN CULTURAL HISTORY
    Eric Avila
AMERICAN FOREIGN RELATIONS
    Andrew Preston
AMERICAN HISTORY
    Paul S. Boyer
AMERICAN IMMIGRATION
    David A. Gerber

AMERICAN INTELLECTUAL
    HISTORY
    Jennifer Ratner-Rosenhagen
THE AMERICAN JUDICIAL
    SYSTEM Charles L. Zelden
AMERICAN LEGAL HISTORY
    G. Edward White
AMERICAN MILITARY HISTORY
    Joseph T. Glatthaar
AMERICAN NAVAL HISTORY
    Craig L. Symonds
AMERICAN POETRY David Caplan
AMERICAN POLITICAL HISTORY
    Donald Critchlow
AMERICAN POLITICAL PARTIES
    AND ELECTIONS L. Sandy Maisel
AMERICAN POLITICS
    Richard M. Valelly
THE AMERICAN PRESIDENCY
    Charles O. Jones
THE AMERICAN REVOLUTION
    Robert J. Allison
AMERICAN SLAVERY
    Heather Andrea Williams
THE AMERICAN SOUTH
    Charles Reagan Wilson
THE AMERICAN WEST Stephen Aron
AMERICAN WOMEN'S HISTORY
    Susan Ware
AMPHIBIANS T. S. Kemp
ANAESTHESIA Aidan O'Donnell
ANALYTIC PHILOSOPHY
    Michael Beaney
ANARCHISM Alex Prichard

For more information visit our website

www.oup.com/vsi/

A. W. Moore

GÖDEL'S THEOREM

A Very Short Introduction

OXFORD
UNIVERSITY PRESS

Great Clarendon Street, Oxford, OX2 6DP,
United Kingdom

Oxford University Press is a department of the University of Oxford.
It furthers the University's objective of excellence in research, scholarship,
and education by publishing worldwide. Oxford is a registered trade mark of
Oxford University Press in the UK and in certain other countries

© A. W. Moore 2022

The moral rights of the author have been asserted

First edition published in 2022

Impression: 1

Published in the United States of America by Oxford University Press
198 Madison Avenue, New York, NY 10016, United States of America

British Library Cataloguing in Publication Data
Data available

Library of Congress Control Number: 2022937768

ISBN 978-0-19-284785-0

Printed in the UK by
Ashford Colour Press Ltd, Gosport, Hampshire

For Ellen and Clare

Contents

Preface

I remember, at an early stage in my academic career, seeking advice from my mentor Bernard Williams about the relative merits for one's career prospects of publishing a book and publishing articles. He made the obvious point: a lot depends on the quality of the book and the quality of the articles. He then contrasted the tiny impact that a book may have with that had by, for example—and here he had a characteristic twinkle in his eye as he quoted the full title—'On Formally Undecidable Propositions of *Principia Mathematica* and Related Systems I'.

He was referring to the 1931 article in which Kurt Gödel published the celebrated theorem that is named after him, an article that had a profound impact. It contained one of those landmark intellectual achievements, like Heisenberg's uncertainty principle or the discovery of the structure of DNA, that fundamentally change people's view of things.

Many people have an inkling of Gödel's theorem but wish they had more. That's what I aim to provide in this book. Fortunately, both the content of the theorem and the outline of its proof can be readily conveyed to anyone with a command of high-school mathematics, which is as much mathematical background as I will be presupposing. I will also say something about the theorem's philosophical implications. And I will attempt to scotch some prevalent misunderstandings of it.

My thanks are due to Latha Menon for encouraging me to write this book; to Ian Sinclair for many hours of discussion about its content, during which his probing questions forced me to think more clearly about how best to convey what I had to say; to two anonymous referees for OUP; and to two former teachers, Neil Sheldon who taught me when I was a sixth former and Timothy Smiley who taught me when I was an undergraduate, for inspiring my interest in these issues.

Note on the Appendix

I have included an Appendix in which I briefly sketch the proof of Gödel's theorem. Different readers will benefit from looking at this at different stages, if at all. For some, it will provide helpful orientation straight away. For most, it will mean little or nothing until the end of Chapter 3, at which stage it may be useful in motivating some of the definitions that I will be providing in Chapter 4. For others, it can serve as a convenient *aide-memoire* after I have proved the theorem in Chapters 5 and 6. For everyone, I hope, it will give sight of the wood that may otherwise not be visible for the trees.

List of illustrations

Chapter 1
What is Gödel's theorem?

On 7 September 1930, towards the end of a conference held in Königsberg on 'Epistemology of the Exact Sciences', Kurt Gödel gave a few details of his famous result. He was 24 years old at the time and had only recently completed his doctorate. He was not one of the conference luminaries. It is unclear how much attention was even paid to what he said. But once his theorem had been published, the following year, under the title 'On Formally Undecidable Propositions of *Principia Mathematica* and Related Systems I'—each component of this title will be explained in due course—it soon began to attract the widespread attention and acclaim it deserved.

Gödel had proved something that challenged prevalent presuppositions about the nature of mathematics and that raised all sorts of philosophical questions. He had also proved something that was of deep mathematical significance in its own right. (The mathematician Roger Penrose has described it as 'the most important theorem in mathematical logic of all time'.) It took one's breath away, not only for the ingenuity of the techniques that were used to establish it but also for its sheer beauty.

Gödel went on to do further significant work, but his name will always be associated primarily with this theorem. He also went on to have severe mental health problems, one notable example being

1. Kurt Gödel.

a paranoid conviction that other people were out to poison him. There are those who have speculated (however implausibly) that these problems were aggravated by the unsettling nature of his work. Fortunately, we don't need to engage in such speculation to acknowledge the power of this work: power not only to unsettle but also to enthral and to inspire.

What then is Gödel's theorem? Here is an initial informal statement of it:

> No axiomatization can determine the whole truth and nothing but the truth concerning arithmetic.

But this calls for elucidation, in particular of both 'axiomatization' and 'arithmetic'. I'll begin with the latter.

Arithmetic

By arithmetic I mean the study of the natural numbers 0, 1, 2, 3,...and the various operations that apply to them, such as addition and multiplication. Sometimes by arithmetic is meant the study of just the positive integers 1, 2, 3, 4,...and the various operations that apply to them. It doesn't make much difference. But there is at least one respect in which the inclusion of 0 is natural (as the label 'natural numbers' suggests). Like the positive integers, 0 can be used to answer questions of the form 'How many...?'. For instance, it can be used to give a correct answer to the quiz question, 'How many clarinet concertos did Beethoven write?'.

What do I mean when I talk of operations 'such as addition and multiplication'? I mean *algorithmic* operations, that is operations that can be carried out by following an algorithm; and by an algorithm I mean a procedure of the sort that a computer could follow—that is, a purely mechanical, step-by-step procedure that doesn't require any insight or ingenuity, that doesn't rely on any random devices such as coin tossing, and that is guaranteed to produce a result after a finite number of steps. There is a procedure of this sort to ascertain the sum of any two natural numbers, another to ascertain their product. (This is why computers can add and multiply.) Exponentiation is another such operation. So is the rather contrived and uninteresting operation which, when applied to any given natural number as input, 14 say, yields the cube of the next largest prime number as output, 17^3 or 4,913 in this case. There are many further operations of this kind, indeed infinitely many.

In fact, however, I needn't have mentioned this larger range of operations. As we'll see in Chapter 6, Gödel's theorem would hold even if arithmetic were understood as the study of the natural numbers and *just* addition and multiplication. This is because all the other algorithmic operations can be defined in terms of those two. (Not that this is as straightforward as it appears. For instance, you might think that exponentiation can readily be defined in terms of multiplication as follows: given any two natural numbers n and m, n to the power of m is n multiplied by n, multiplied by n, multiplied by ... n, where the n occurs m times. But that idea of 'occurring m times' must in turn be suitably defined. This turns out to be not at all straightforward.) The fact that all the other algorithmic operations can be defined in terms of addition and multiplication is doubly striking: it shows both how much can be done with such limited resources and how little it takes for Gödel's theorem to apply.

Axiomatization

An axiomatization is a finite stock of basic principles and rules. This is connected with what we have just been discussing, since paradigmatic examples include the basic principles and rules that appear in an algorithm. For instance, a basic principle that might appear in an algorithm for multiplication is the following: given any natural number n, n multiplied by 0 is 0. And a basic *rule* that might appear in an algorithm for multiplication is the following: given any principle of the kind just mentioned, there is licence to derive any particular instance of it, which, in this case, means there is licence to conclude that 7 (say) multiplied by 0 is 0. Such principles and rules provide the elementary steps that we must take to carry out more complicated calculations, say to calculate the product of 827 and 240.

For these purposes, a principle or a rule counts as 'basic' when it is accepted without being derived from anything else. Whether a basic principle or rule *deserves* to be accepted without being derived from anything else is a further question. Its basicness is not compromised by the fact, if it is a fact, that it can be called into question—nor even by the fact, if it is a fact, that it can be shown not to be correct.

In practice, this is liable not to be an issue (although it sometimes is, as we'll see). Most of the principles and rules that we accept without deriving them from anything else are not just correct, but undeniably correct, and we find ourselves simply taking them for granted. Even then, however, there remain difficult philosophical questions about what is at stake. Does our confidence in these principles and rules betray sensitivity to how things are, independently of us? Or are we in effect conferring their correctness on them, as when we say that the bishop in chess moves diagonally? Such questions, fascinating though they are, need not concern us here.

Note that, in order for a basic principle or rule to be of relevance to arithmetic, it doesn't need to be peculiarly arithmetical. The rule that gives us licence to derive any particular instance of any given general principle isn't. *That* rule could equally serve in an axiomatization designed to determine the truth concerning physics, say, or human rights. Basic principles and rules that don't peculiarly concern any one subject are often said to belong to *logic*. It would be by appeal to logic, so understood, that we would deduce the impossibility of mating a lone king with a king and a knight, given the rules of chess. The rules of chess are not themselves a matter of logic: they are stipulations that serve to define the game. But it *is* a matter of logic that, *given* these rules, there can be no such mate.

What exactly marks the distinction between what belongs to logic and what does not, or even on which side certain important principles and rules lie, are matters of some controversy. For our purposes, the loose, informal characterization of the distinction suggested above will suffice: logic pertains to all subjects. I will presuppose this distinction in much of what follows, highlighting points at which the controversy matters.

Proof

Any given axiomatization will provide canons of proof. A proof of a statement, relative to the axiomatization, will be a derivation of that statement from those basic principles, using those basic rules. Thus a decent axiomatization of arithmetic should contain basic principles and rules that enable us to calculate the product of 827 and 240—in other words, to prove that $827 \times 240 = 198{,}480$. It should also contain basic principles and rules that enable us to prove much more interesting results, such as that there are infinitely many primes.

That there are infinitely many primes can be proved as follows:

Let $p_1, p_2, \ldots, p_n$ be any finite list of primes. Let q be their product. Consider $q + 1$. Either $q + 1$ is prime or it is not. If it is, then it is a prime that is not in the original list, because it is larger than any of them. If it is not, then it has a prime factor that is not in the original list, because none of them divides it. (All of them divide q.) Either way, the original list excludes at least one prime. So there cannot be only finitely many primes.

The proof given here is modelled on one that Euclid gave around 300 BC. And it is perfectly acceptable. But the relevant axiomatization has not been made explicit. A more formal proof of the result would need to indicate *exactly* what basic principles and rules were being used in it. And indeed one of the most interesting questions about any given proof may well be exactly what basic principles and rules *are* being used in it. That question may be hard to answer even when the acceptability of the proof is not in doubt.

Note the relativization incidentally. This is very important. We will never ask whether something is a proof or not *tout court*. The question will always be whether something is a proof or not *relative to a given axiomatization*. What is provable relative to one axiomatization may not be provable relative to another. That is, what is provable using one finite stock of basic principles and rules may not be provable using another: the former may include a principle or a rule that the latter lacks. (This reflects the intuitive point that what one counts as a demonstration of something is determined, in part, by what one is prepared to take for granted.)

Arithmetical truths and falsehoods

My initial informal statement of Gödel's theorem was in terms of the concept of truth. Here is another, equivalent statement in terms of the concept of *truths* in the plural (and its complement, the concept of falsehoods):

Any axiomatization of arithmetic must either be incomplete, that is fail to capture some arithmetical truths, or be unsound, that is capture some arithmetical falsehoods.

Here are some examples of what I mean by arithmetical truths and falsehoods:

- $7 + 5 = 12$. (This is a familiar truth.)
- $7 + 5 = 13$. (This is a corresponding falsehood.)

- There are no three natural numbers, k, l, and m, such that $k + l$, $k + m$, and $l + m$ are all odd. (This is a truth that differs from either of the two previous statements in being a generalization rather than a statement about specific natural numbers. I leave its proof as an exercise.)

- There are infinitely many primes. (This is the truth whose proof I gave above.)

- Every natural number is the sum of three squares, as, for example, $3 = 1 + 1 + 1$, or $20 = 16 + 4 + 0$. (This is a falsehood, despite these examples and despite infinitely many other examples. It is not true of 7, for instance.)

- Every natural number is the sum of *four* squares, as, for example, $7 = 4 + 1 + 1 + 1$, or $21 = 16 + 4 + 1 + 0$. (This is a truth that was proved in 1770 by Joseph Lagrange using techniques that are far beyond the scope of this book.)

- There are no four positive natural numbers k, l, m, and n, such that n is greater than 2 and such that $k^n + l^n = m^n$. (This is a truth known as Fermat's last theorem. It is named after Pierre de Fermat, who, in the 17th century, wrote in the margin of his copy of Diophantus' *Arithmetica* that he had a proof of it. The consensus nowadays is that he did not, and that it was not proved until the mid-1990s, by Andrew Wiles, using techniques that are not only far beyond the scope of this book but far beyond many professional mathematicians.)

- Every even number greater than 2 is the sum of two primes, as, for example, $8 = 5 + 3$, or $22 = 11 + 11$. (This is a statement known as Goldbach's conjecture. It is named after Christian Goldbach, who mooted it in the 18th century. I can't say whether it is a truth or a falsehood. No-one knows. Or at least, no-one knows at the time of my writing this. No-one has discovered a counterexample to the conjecture, but neither has anyone proved it.)

I am assuming, incidentally, at least for the time being, that standard arithmetical terminology has a clear intended meaning and that every statement that uses only such terminology can unproblematically be regarded as either true or false. This assumption will come under scrutiny later. It is worth noting already, however, that there are reasons to be cautious about it, and not just reasons arising from philosophical scruples. What exactly counts as 'standard arithmetical terminology'? (The statements above are couched in a mixture of arithmetical notation and ordinary English.) Does such terminology include symbols like '÷' and '√'—the division symbol and the square root symbol, respectively? If so, then won't it also include expressions such as '1 ÷ 0' and '√2' which lack (arithmetical) meaning and which thereby render any arithmetical statements in which they occur neither true nor false? Suffice to say, at this stage, we'll need to be alert to such dangers when they arise. But it is also worth emphasizing that, despite the informal statements of Gödel's theorem that I have given so far, the theorem *can* be stated and proved in a way that bypasses the concept of truth altogether, as we'll see in due course.

The upshot at this stage, then, is that, according to Gödel's theorem, no finite stock of basic principles and rules enables us to prove every arithmetical truth unless it also enables us to prove some arithmetical falsehoods.

The significance of the word 'finite' and the significance of the qualification 'unless it also enables us to prove some arithmetical falsehoods' both deserve to be emphasized. Without the former, we'd have no principled way of resisting the suggestion that each of the infinitely many truths of arithmetic, however unobvious, should serve as a basic principle in its own right, enabling us to use it as a one-line proof of itself. (Note, however, that there is a crucial subtlety concerning just what the finitude of a stock of basic principles and rules amounts to here; we'll return to this in Chapter 4.) As far as the latter is concerned, consider a basic rule that gives us licence, at any point, in any proof, to derive anything at all. Clearly this (absurd) rule, although it enables us to prove every arithmetical truth, does so only at the unacceptable price of enabling us to prove every arithmetical falsehood too.

Misconceptions of Gödel's theorem

The informal statements of Gödel's theorem that I have given so far are already enough to scotch two common misconceptions of the theorem.

First, Gödel's theorem does not entail that some arithmetical truths are unprovable. It does not entail this even when proof is understood in the axiomatic way in which it is being understood here and even when attention is restricted to axiomatizations that are sound, that is axiomatizations that capture nothing but truths. Recall the relativization in the concept of proof that I emphasized earlier. An unprovable truth would be a truth that could not be proved *in any sound axiomatization whatsoever*. Gödel's theorem gives us no reason to believe in any such thing. Apart from anything else, there is always the possibility that I flagged above of a truth's serving as a basic principle in its own right, enabling us to use it as a one-line proof of itself. We can put it this way: the fact that any sound axiomatization fails to capture some arithmetical truths does not mean that some arithmetical truths fail to be captured by any sound axiomatization (any more than the fact that no single camera in a CCTV system covers the entire site means that there are some parts of the site that are not covered by any of the cameras).

Second, Gödel's theorem does not preclude axiomatizations that determine the whole truth and nothing but the truth concerning other subjects: it specifically concerns arithmetic. It does not even preclude axiomatizations that determine the whole truth and nothing but the truth concerning other branches of mathematics. A notable example of a branch of mathematics for which there *is* such an axiomatization is analysis. Analysis is the study of the real numbers—that is, the negative and non-negative integers together with all the numbers in between, such as $1\frac{1}{2}$, $-1\frac{1}{2}$, $\sqrt{2}$, and π—and the same two operations as are integral to arithmetic: addition and multiplication.

This is notable not least because it is so counterintuitive. Analysis *appears* to be arithmetic and more besides. (Natural numbers are themselves real numbers.) So you might well have thought that any axiomatization that determines the whole truth and nothing but the truth concerning analysis cannot help but determine the whole truth and nothing but the truth concerning arithmetic as well. And indeed the axiomatization of analysis does enable us to prove truths about specific natural numbers, such as that $7 + 5 = 12$. The point, however, is that it does not enable us to prove truths about the natural numbers as a whole. It does not enable us to do this because, although there is provision in its language for us to refer to specific natural numbers, there is no provision in its language for us to refer to the natural numbers *as such—as* the natural numbers. It is as if we had a language for discussing heavenly bodies that included the names 'Mercury', 'Venus', 'Mars', and so forth, but not the word 'planet'.

There are other misconceptions associated with Gödel's theorem whose allure is harder to appreciate at this stage, although I can already say something about them. For example, it is often claimed that Gödel showed that there are arithmetical statements that assert their own unprovability using this or that axiomatization. Why anyone would make such a claim should be clearer in due course. And in fact we *can* eventually say something of this sort. But we have to work extraordinarily hard to earn the right to say it. Until we have done the work, and in particular until we have clarified what exactly we mean, then this too counts as a misconception. For the time being, we are better off just adhering to the commonsense view that, in so far as an arithmetical statement can be said to 'assert' anything at all, then what it asserts is something about natural numbers.

Other misconceptions are much wilder. There are those who see in Gödel's theorem, with its nod towards what is infinite and resists being captured by what is finite, justification for a belief in

God. Less wild, but still in my view wild, are these two claims, made by Rudy Rucker and David Foster Wallace, respectively:

> [I]f Gödel's theorem tells us anything, it is this: Man will never know the final secret of the universe.

> After Gödel, the idea that mathematics was not just a language of God but language we could decode to understand the universe and understand everything—that just doesn't work any more. It's part of the great postmodern uncertainty that we live in.

There is simply no interesting connection between Gödel's theorem and the ideas being canvassed in these claims. In the case of Wallace's claim, for instance, we can ask what it is he thinks worked, or looked as though it worked, *pre*-Gödel. Whatever plausible answer he gives, it won't have anything to do with the axiomatization of arithmetic.

Fortunately, Gödel's theorem loses none of its interest or importance when these and other misconceptions are scotched.

Gödel's two theorems

I have been talking, as many people do, about 'Gödel's theorem' in the singular. We'll see in Chapter 6 that this is somewhat misleading. This is not just because Gödel also proved theorems about quite different matters, including, in his doctoral dissertation, an important theorem about logic that will be relevant in Chapter 4. Even when his theorem about the axiomatization of arithmetic has been distinguished from these others, typically as his 'incompleteness' theorem, there is a further distinction that is often drawn between his 'first' incompleteness theorem and his 'second'. This will be clarified in Chapter 6.

It gives me an early excuse, however, to explain one part of the title of Gödel's article—the curious 'I' at the end. Gödel originally intended his article as the first of a pair. In the published article he proved his first incompleteness theorem in detail, with reference to a specific axiomatization. He also sketched a proof of the second. In the sequel, he had intended to show how the first incompleteness theorem extends to other axiomatizations, and then to prove the second in detail. As it turned out, he never published the sequel. This was because there was no need—his results were accepted straight away; there was already sufficient consensus about what the sequel would have contained.

Chapter 2
Axiomatization: its appeal and demands

One aim of this chapter will be to highlight the appeal of what Gödel's theorem precludes—this being a large part of the significance of the theorem and the reason why that significance appears, initially at least, in a negative light. (In later chapters its significance will appear in a more positive light.)

The general appeal of axiomatization

There is something very enticing about the idea that the entire (infinite) truth concerning some subject can be captured in a (finite) axiomatization. If it can, then that will not only help us to make sense of the subject; it will also help us to make sense of how we make sense of it.

We find evidence of this enticement in all sorts of contexts. In the Gospel of St Matthew, Jesus tells a lawyer that the two greatest commandments are to love God unconditionally and to love one's neighbour as oneself; then he adds, 'On these two commandments hang all the law and the prophets'. The philosopher Benedictus de Spinoza, writing in the 17th century, presented his masterpiece *Ethics* in an axiomatic form, deriving his conclusions from a stock of such basic principles as 'Man thinks'. A little later, the scientist Isaac Newton also used an axiomatic form to present his own masterpiece *Mathematical Principles of Natural Philosophy*, the first of his basic principles being that a body continues in a state of rest or of uniform motion in a straight line unless it is compelled to change that state by forces acting upon it. The American Declaration of Independence, written in 1776, famously begins—after a preambulatory paragraph—'We hold these truths to be self-evident', and then lists basic principles that are intended to justify what follows, the first being that all men are created equal. And work has recently been done to show that special relativity can be based on a finite number of basic principles, one of which is that, for any inertial observer, the speed of light is the same everywhere and in every direction. In each of these examples an attempt is being made, however formally or informally, to capture an unlimited wealth of knowledge or wisdom in a manageably small stock of basic principles and rules that can be accepted as such.

The mathematical appeal of axiomatization

These examples concern non-mathematical subjects. In the case of a mathematical subject, appeal to such an axiomatization may even appear to be the *only* way in which we can make sense of it—and, therefore, the only way in which we can make sense of how we make sense of it. For how else can we make sense of a mathematical subject, if not by seeing what it takes to prove truths concerning that subject? And what can constitute our canons of proof, if not some finite stock of basic principles and rules that we have at our disposal? What sense do we even have of what a truth concerning a mathematical subject is, if it's not a statement that we have such means, at least in principle, to prove?

Suppose, by way of illustration, that someone were to claim the following:

> Division by 0 in an arithmetical context is legitimate. There is an answer to the question, 'What is $1 \div 0$?' It is just that we can't tell what the answer is. *We* can only tell that $l \div m = n$

15

when the algorithm for multiplication enables us to prove that $l = m \times n$. But that doesn't help in this case. We'll never know whether $1 \div 0 = 0$, or whether $1 \div 0 = 1$, or whether $1 \div 0 = 2$, or...

This would be absurd. For a more subtle—and more contentious—example, reconsider Goldbach's conjecture, that every even number greater than 2 is the sum of two primes, which I used as one of my examples in Chapter 1 and which, as I noted, no-one has yet either proved or disproved. There are those who think that it would be equally absurd to claim that this conjecture might be true even without there being suitable canons of proof that we could use to tell that it was true. On this view, it would not be possible for the conjecture to be true as a matter of brute fact—as some kind of 'infinite coincidence' involving each and every even number greater than 2. There are some extremely difficult philosophical issues at stake here, and they will be relevant to some of the subsequent discussion too. Fortunately, however, we do not need to resolve them to acknowledge further indication here of the allure of axiomatization.

A related issue is how any of us knows what any mathematical vocabulary means. In particular, how does any of us know what any arithmetical vocabulary means? Consider the very expression 'natural number'. Imagine someone who did *not* know what this meant. You might give them an explicit definition. You might define 'natural number' as 'non-negative integer'. This would be to no avail, however, if they were as unfamiliar with that mathematical terminology as with the expression 'natural number' itself. You might then do the sort of thing I did in Chapter 1 and say, 'The natural numbers are the numbers 0, 1, 2, 3, and so on', perhaps adding, as I also did in Chapter 1, that they are numbers that can be used to answer questions of the form 'How many...?' In practice, this would no doubt work. Strictly speaking, however, there would be some slack here that still needed to be negotiated. Mathematicians nowadays are comfortable with the idea of infinite numbers, and such numbers can also be used to answer questions of the form 'How many...?' (How many points are there on a line?) You wouldn't have said anything so far to prevent your use of 'and so on', in the expression '0, 1, 2, 3, and so on', from extending as far as such infinite numbers, nor therefore to preclude their classification as 'natural numbers'. You could add, 'There are no infinite natural numbers'. But now you would need to say what you meant by 'infinite', to indicate precisely what you were ruling out. In practice, none of this would be a problem. But it *would* be a problem in theory. And the fact that it would be a problem in theory means that there is a question about why it wouldn't be a problem in practice. How has any of us actually managed to acquire an understanding of the expression 'natural number'? How has any of us got into a position to see what's going on here?

We must have been exposed to *something* that has conferred our understanding of arithmetical vocabulary on us. It is tempting to conclude that there must be some finite list of instructions for how to use that vocabulary which goes beyond these appeals to questions of the form 'How many...?' and which each of us has in some way or other assimilated. But what form could this finite list of instructions take if not something equivalent to an axiomatization that captures all and only the truths of arithmetic? Yet this is precisely what Gödel's theorem precludes.

We must think again. We must reconsider what is required to understand arithmetical vocabulary. (This is an issue we'll come back to in Chapter 8.) But here is another indication of why Gödel's theorem creates some discomfort for us.

Euclid's *Elements*: a (flawed) paradigm

Around 300 BC Euclid compiled his *Elements*, a mathematical work consisting of thirteen books. This work is now generally agreed to have been based on the work of earlier mathematicians.

Whether it was the product of one person or of many, it was an extraordinary achievement. It exhibited dazzling power, precision, and beauty. Until quite recently, it was one of those works that any self-respecting educated person would have read.

Especially noteworthy, and especially relevant to the current discussion, is the axiomatization of geometry included in it. (By 'geometry' I mean the study of points, lines, and shapes in a plane.) This is a paradigm of what we have been talking about. Precisely what it is is a finite stock of basic principles and rules intended to determine the whole truth and nothing but the truth concerning geometry. The basic principles in question comprise what Euclid calls 'postulates' and 'common notions'—five of each—along with twenty-three definitions of key terms. (The distinction between the postulates and the common notions is a variation on the distinction I drew in Chapter 1 between basic principles that are non-logical and basic principles that are logical. The postulates peculiarly concern geometry; the common notions might appear in axiomatizations of other subjects too.) Among the postulates, there is one according to which, given any two points p_1 and p_2, a straight line can be drawn from p_1 to p_2; and there is another according to which, given any point p and given any distance d, a circle can be drawn with p as its centre and d as its radius. Among the common notions there is one according to which 'things which are equal to the same thing are also equal to each other', and another according to which 'the whole is greater than the part'. And among the definitions, there is the standard definition of *parallel lines* as two straight lines in a given plane that never meet, however far they are extended in either direction.

To repeat, this axiomatization of geometry is a paradigm of what we have been talking about. Euclid's basic principles can be used to derive more and more sophisticated geometrical results, including Pythagoras' famous theorem that the square on the hypotenuse of a right-angled triangle is equal to the sum of the

18

squares on the other two sides. There is an infinite wealth of learning captured in this finite stock of basic principles.

Even so, the paradigm is flawed. It does not satisfy modern standards of rigour. A (sound and complete) axiomatization of Greek geometry that satisfies these standards *has* been devised, but Euclid's own efforts needed to be supplemented and modified to achieve this.

The first and most glaring problem with Euclid's axiomatization, from a modern point of view, is that it does not include a specification of the linguistic resources on which it will draw. Despite his definitions, Euclid never indicates the full range of the vocabulary that will be used, or how that vocabulary is to be combined to form the statements whose truth or falsity is at stake. He does not provide the first thing that is required of a mathematical axiomatization from a modern point of view.

Here it is noteworthy that some of Euclid's postulates, in his own formulation, don't even qualify as statements. The postulate that a straight line can be drawn between any two points, for example, is not formulated in that way by Euclid. It is formulated as follows: 'To draw a straight line from any point to any point'. That is not a statement. It is an infinitive verb phrase, and, as such, neither true nor false. We can understand why Euclid formulates the postulate in this way. He is registering the fact that it is concerned with a construction. And I am not criticizing him for doing this. The fact remains that he is thereby violating modern scruples. On a modern conception, an axiomatization is a means of deriving certain *statements* from certain other *statements*. What is at stake is what must be true (and therefore capable of being true) if the basic principles themselves are true.

Why does specifying the statements whose truth or falsity is at stake matter? Well, recall the axiomatization of analysis that I mentioned in Chapter 1. Given what belongs to the *language* of analysis, this axiomatization is complete: it captures all the true statements in that language. It does not, however, capture all the true statements of arithmetic, even though its subject matter includes the natural numbers and the operations of addition and multiplication. This is because there are some true statements of arithmetic, and in particular there are some true generalizations about the natural numbers, that don't belong to its language. If resources were added to the language so that these generalizations did belong to it, the axiomatization would no longer be complete. The truth or falsity of new statements would be at stake; and among them would be some truths that the axiomatization failed to capture. So it matters.

In particular, it matters in Euclid's own case. To see why, note that Euclid's axiomatization does not enable us to prove that a straight line between two points is the shortest line between them. It enables us to prove that a straight line between two points is shorter than any dog-leg between them (made up of two straight lines): this is sometimes known as the triangle inequality theorem. But it does not enable us to prove that a straight line between two points is shorter than any curved line between them. Question: does this mean that Euclid's axiomatization is incomplete? Answer: *it depends what the relevant linguistic resources are*. If there is no provision in the language of Euclidean geometry for comparing the lengths of straight lines and curved lines, then it does *not* mean that Euclid's axiomatization is incomplete. To think it does would be like thinking that Euclid's axiomatization is incomplete because it doesn't enable us to prove that water contains oxygen. The problem, however, is that we have not been told whether there is provision in the language for making such comparisons. I am not saying that there should be or that there shouldn't be. I am saying that it should be made clear whether there is.

Another flaw in Euclid's axiomatization, from a modern point of view, is that, although it includes basic principles, it doesn't include any basic rules for deriving anything from them. The rules that are actually used in the *Elements* are a matter of logic, and Euclid may have thought they went without saying. From a modern point of view, however, nothing goes without saying.

Yet another flaw in the axiomatization is that not only does it require us to take for granted some logic that is never made explicit, it requires us to take for granted some *geometry* that is never made explicit. Admittedly, we have to look very carefully to see why. At first sight, every single one of Euclid's proofs is beyond reproach in this respect. Indeed one of the marvels of the *Elements* is how assiduously Euclid strives to ensure that nothing geometrical is taken for granted in any of his proofs—other than the basic principles themselves and what has already been derived from them. In fact, however, he falls at the very first hurdle, with his first proof.

That proof is supposed to show that, given any finite straight line, an equilateral triangle can be drawn with that line as one of its sides. The proof proceeds as follows.

Let AB be the given finite straight line (see Figure 2). A circle can be drawn with centre A and radius equal to the length of AB. Another circle can be drawn with centre B and radius equal to the length of AB. Let C be one of the points at which the two circles intersect, and let straight lines be drawn between A and C, and between B and C. Then the length of CA equals that of AB, since they are both equal to the radius of the first circle, while the length of CB equals that of AB, since they are both equal to the radius of the second circle. And the length of CA equals that of CB, since they are both equal to the length of AB. So the triangle ABC is the required equilateral triangle.

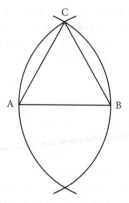

2. Euclid's proof that an equilateral triangle can be drawn on any given finite straight line.

This proof looks impeccable. In fact, however, there is no guarantee in Euclid's basic principles that the two circles intersect at any point. If we look at Figure 2, it will strike us as obvious that they do. But that counts for nothing. Unless whatever strikes us as obvious when we look at Figure 2 is already captured by Euclid's basic principles, it should be incorporated into a new one.

The point here is not to deprecate anything that Euclid did. On the contrary, the point is to emphasize how hard it is to provide an unimpeachable axiomatization, which helps us to appreciate all the more what Euclid achieved in coming as close as he did. The upshot of this part of the discussion is that Euclid's paradigm, while it certainly serves as a danger signal of how easy it is to get things wrong in an axiomatization, also serves as an inspiration for trying to get them right.

Independence and consistency

There is a further important issue on which Euclid's *Elements* bears. It concerns the fifth of his postulates. Let us call a straight line that extends indefinitely far in both directions STRAIGHT.

Then Euclid's fifth postulate, in a formulation different from his but equivalent to it, is as follows:

Euclid's Fifth Postulate: Given any STRAIGHT line *l* and given any point *p* not on *l*, there is one and only one STRAIGHT line that passes through *p* and that is parallel to *l*. (See Figure 3.)

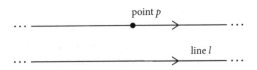

3. Euclid's fifth postulate.

This basic principle is less self-evident than the others. For centuries mathematicians wondered whether it can be derived from the others and can therefore be dispensed with. (Their primary question was not, 'Do we need any more than this?'—the question of completeness—but rather, 'Do we need as much as this?')

When it was eventually shown, in the early 19th century, that Euclid's fifth postulate *can't* be derived from his other basic principles, this was tantamount to showing that there is some rival to the fifth postulate, that's to say some postulate whereby there are sometimes two or more such parallel lines or none at all, which the other basic principles can't be used to rule out. Let us call such a rival postulate a *non-Euclidean* postulate; and let us call an axiomatization that differs from Euclid's only in including a non-Euclidean postulate in place of his fifth postulate a *non-Euclidean* axiomatization. Then this in turn was tantamount to showing that there is some non-Euclidean axiomatization that can be said to make its own sense. But what does 'make its own sense' mean here?

It signals a kind of internal coherence that an axiomatization may enjoy even if it captures some falsehoods. (This is a bit like the kind of internal coherence that a story about a princess on the run

23

from her stepmother and living with seven dwarfs may enjoy even if no such thing has ever happened or ever will happen.) The kind of internal coherence in question is of great significance, not only to mathematics in general but also to Gödel's theorem in particular. In Chapter 4, I will give a more precise account of what it comes to. But already I can give an indication.

Let us focus on one particular non-Euclidean axiomatization **NA** which includes the following non-Euclidean postulate *NP*:

> (*NP*) Given any STRAIGHT line *l* and given any point *p* not on *l*, there is more than one STRAIGHT line that passes through *p* and that is parallel to *l*.

As I have already indicated, **NA** may enjoy internal coherence of the relevant kind even if it captures some falsehoods, and in particular even if *NP* itself is false. (I am leaving open for the time being the question whether *NP*, or any other non-Euclidean postulate, *is* false. I have not committed myself to the soundness of Euclid's own axiomatization.) For **NA** *not* to enjoy internal coherence of the relevant kind would be for it not just to capture some falsehoods, but to capture some falsehoods *purely as a matter of logic*. Suppose that the rest of Euclid's basic principles *had* enabled us to prove his fifth postulate. Then **NA** would have been subject to precisely this criticism. We could have used it to prove both that there is only ever one such parallel line, in accord with the fifth postulate, and that there is *not* only ever one such parallel line, in accord with *NP*—and one of these must be false, on purely logical grounds. As it is, **NA** is immune to any such criticism. (I will sketch a proof of this shortly.) The sense that **NA** would have failed to make, if it had involved such a contradiction, and the sense that it actually does make, by not involving such a contradiction, is *logical* sense. **NA** would have failed to be, but actually is, *consistent*.

In Chapter 1, I drew a distinction between the logical and the non-logical with respect to basic principles and rules. We can draw a similar distinction with respect to vocabulary. Logical words and phrases are used in connection with any subject: examples are 'not', 'and', 'every', and 'the same as'. Non-logical words and phrases are used in connection with a particular subject: examples are 'point' and 'line', which are used specifically in connection with geometry. Given this distinction, we can cast consistency and its opposite, inconsistency, in the following terms. An axiomatization is inconsistent if we can tell that it captures some falsehoods just by looking at the logical vocabulary involved; it is consistent otherwise.

As a further illustration, let 'ryma' and 'zodlite' be two (non-logical) words whose meaning has not yet been disclosed to us; and let A_i and A_c be two axiomatizations involving these two words. Suppose that A_i enables us to prove both 'Some rymas are zodlite' and 'No rymas are zodlite'. Then A_i is inconsistent. The sheer meaning of the logical words 'some', 'no', and 'are' ensures that these statements can't both be true—irrespective of what 'ryma' and 'zodlite' mean. Suppose, by contrast, that A_c is consistent, enabling us to prove 'Some rymas are zodlite', but not 'No rymas are zodlite'. Then there is no such obstacle to our thinking that all the statements that A_c enables us to prove are true, in other words that A_c itself is sound.

Of course, whether A_c *is* sound (unlike whether A_i is sound) depends on what 'ryma' and 'zodlite' actually mean. Until the meaning of these two words has been disclosed to us, we cannot know whether A_c is sound—except, perhaps, by appeal to some authority. We may however be in a position to know a related hypothetical. There may be a candidate meaning for 'ryma' and 'zodlite' such that we can see that, *if* they had *this* meaning, then A_c would be sound, because then all its basic principles and rules would be correct. And if we are in a position to know a hypothetical such as this, then we are also in a position to know, non-hypothetically, that A_c is consistent—despite the fact that we do not know what 'ryma' and 'zodlite' actually mean. For unless A_c were consistent, it could not be sound, whatever they meant.

Herein lies the clue as to how we show that the non-Euclidean axiomatization **NA** is consistent, and, as a result, that Euclid's fifth postulate cannot be derived from the rest of his basic principles. We first of all prescind from the actual meaning of the non-logical vocabulary in **NA**—'point', 'line', and so on. That is, we treat this vocabulary as though it consisted of alien terms like 'ryma' and 'zodlite' whose meaning has not yet been disclosed to us. In particular, we think of the non-Euclidean postulate *NP* as something like *NP**:

> (*NP**) Given any zodlite ryma r and given any bumid b that does not fruminize r, there is more than one zodlite ryma that padveates b and that bellerates r.

We then find a set of candidate meanings for these alien terms, or what is standardly called an *interpretation* of them—possibly of a geometrical kind, possibly having nothing at all to do with geometry—on which all of **NA**'s basic principles are true. We then conclude, by the argument above, that **NA** is consistent.

As an indication of the *sort* of thing that we are looking for, here is a non-geometrical interpretation of the alien terms in *NP** on which *NP** is true:

'zodlite'	is interpreted as	'English'
'ryma'	is interpreted as	'word'
'bumid'	is interpreted as	'letter'
'fruminize'	is interpreted as	'occur in'
'padveates'	is interpreted as	'contains'
'bellerates'	is interpreted as	'has at least one letter in common with'.

Not that this interpretation is itself what we are looking for. It cannot be extended to an interpretation of the alien terms in Euclid's other basic principles on which they are also true.

But now consider the following geometrical interpretation of the alien terms in *NP**. Imagine a horizontal STRAIGHT line *h*. And call any line above *h* 'zodlite' if it is either perpendicular to *h*, continuing indefinitely far upwards, or a semicircle whose centre lies on *h*. (See Figure 4.) With 'zodlite' interpreted in this way, the remaining alien terms in *NP** are interpreted as follows:

'ryma'	is interpreted as	'line above *h*'
'bumid'	is interpreted as	'point above *h*'
'fruminize'	is interpreted as	'lie on'
'padveates'	is interpreted as	'passes through'
'bellerates'	is interpreted as	'has no point in common with'.

This is another interpretation on which *NP** is true. But, unlike the previous interpretation, it *can* be extended to an interpretation of the alien terms in Euclid's other basic principles on which they are also true (though it is unfortunately beyond the scope of this book to show how). So this is what we are looking for. We are now in a position to conclude that **NA** is consistent.

Note, however, that this interpretation involves a *Euclidean* understanding of the region above *h*. It is as though we have turned Euclidean geometry against itself. We have used Euclidean geometry to show how one of its rivals makes as much logical

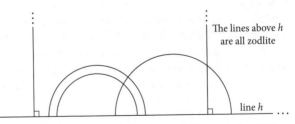

The lines above *h* are all zodlite

line *h*

4. A geometrical interpretation of the non-Euclidean postulate *NP**.

sense as it does. If Euclid's axiomatization makes *no* logical sense, in other words if it is not itself consistent—which we have been assuming it is—then our argument for the consistency of **NA** is worthless.

How then are we to justify our assumption that Euclid's axiomatization is consistent? That's the cue for the next section.

Justifying the assumption of consistency

We might try to justify our assumption that Euclid's axiomatization is consistent by appeal to its soundness. But is it sound? Are all its basic principles true?

That raises the question of what 'true' means here. So far in this book I have made rather free and easy use of the concept of truth. The time has come to think a little more critically about it.

There is one popular view according to which, in a *mathematical* context, sheer consistency is itself a guarantor of truth: soundness and consistency are the same thing. The idea is as follows. Given any consistent axiomatization, there are certain conditions that an interpretation of its non-logical vocabulary must meet to render it sound. But it is a peculiarity of *mathematical* vocabulary (on this view) that there is nothing more to its meaning than satisfying these conditions. So a consistent mathematical axiomatization can be thought of as providing a true description of the corresponding conditions. This allows for the possibility that both Euclid's axiomatization and **NA** are sound, the one because it provides a true description of the conditions that an interpretation of 'point', 'line', and so on, must meet to count as Euclidean; the other because it provides a true description of the conditions that an interpretation of these terms must meet to count as relevantly non-Euclidean. But if *that's* what it takes for Euclid's axiomatization to be sound, then any appeal to its soundness, to justify our assumption that it is consistent, is straightforwardly question-begging.

Very well; but what if this view of mathematics is wrong? What if these geometrical axiomatizations are intended to provide descriptions, not of abstract conditions of this sort, but of *real physical space*? (This is surely how Euclid himself intended his axiomatization.) Well, if they are, and if Euclid's axiomatization admits of consistent non-Euclidean rivals such as **NA**, then hadn't we better keep an open mind concerning whether it is one of these rivals, rather than Euclid's axiomatization, that provides the true description? Perhaps real physical space is non-Euclidean?

And in fact, given the great discoveries of 20th-century physics, it transpires that real physical space *is* non-Euclidean (although in a way that differs from that depicted by **NA** and for reasons that again, unfortunately, lie beyond the scope of this book). For the hundreds of years that people had been studying Euclid's *Elements* and working with it, they had been oblivious to what most of them would have regarded as the really damning objection to it, which is that much of it, when taken as providing a description of real physical space, was false.

So the question of how we are to justify our assumption that Euclid's axiomatization is consistent remains a very pertinent one. And even if there is some mathematical demonstration of its consistency, whether of the same kind as the demonstration of **NA**'s consistency that we considered in the previous section or of some entirely different kind, it looks as though that will only push the problem back. How are we to justify our assumption that the basic principles and rules at work in this demonstration are themselves consistent? Surely, whatever mathematical means we use to show that an axiomatization is consistent, the sheer fact that this is itself an exercise in mathematics means that the same issue will arise with respect to those means. The best we can hope for, it seems, in demonstrating the consistency of an axiomatization, is to use another axiomatization whose consistency (for whatever reason) is less open to doubt.

For the time being, I am couching these issues in rather vague and intuitive terms. But they cut deep. And this very discussion is a foretaste of something fundamental to come later. Be that as it may, one thing that has emerged in the last two sections of this chapter is that there is a concept that, in its own way, is even more basic than the concepts of truth and falsity—that of consistency.

Chapter 3
Historical background

The remarkable efforts of Euclid that we considered in Chapter 2 revealed that the allure of axiomatization was felt nearly two and a half millennia ago. Nevertheless, it wasn't until the 19th century that any serious attempt was made to do for arithmetic, or indeed for any other branch of mathematics, the kind of thing that Euclid had done for geometry. This serves as an important corrective to any temptation we might feel to exaggerate the lessons of Chapter 2. Let us not think that all work in mathematics has to be understood with explicit reference to some axiomatization, either actual or sought after. People were engaged in arithmetic for thousands of years before any attempt was made to specify basic principles and rules for it.

To be sure, basic principles and rules had been specified for certain restricted parts of arithmetic. In particular, there were the basic principles and rules that appear in standard algorithms for addition and multiplication. Children had long been taught these at school, just as they still are; people had long made use of them in their daily lives, just as they still do. But, before the 19th century, whenever a proof of some general arithmetical result had been accepted—such as Lagrange's proof, to which I have previously referred, that every natural number is the sum of four squares—it had been accepted without the benefit of any pre-given canons of proof to ratify it. Some proofs just seemed compelling to mathematicians. No attempt had been made to codify what made them seem so.

One of the eventual spurs to remedy this was a self-consciousness born of the fact that some of what had seemed compelling turned out, on closer inspection, not to be. Indeed some of what had seemed compelling even at the most fundamental level had turned out, on closer inspection, not to be. We saw this when considering the flaw in Euclid's very first proof. Similar problems afflicted arithmetic. For instance, around the turn of the 17th and 18th centuries, the mathematician G.W. Leibniz had proffered what seemed an incontrovertible proof that 2 + 2 = 4. This proof had begun with a definition of 2 as 1 + 1, of 3 as 2 + 1, and of 4 as 3 + 1, and had then proceeded as follows.

> 2 + 2 = 2 + 1 + 1 (by the definition of 2). And 2 + 1 + 1 = 3 + 1 (by the definition of 3). And 3 + 1 = 4 (by the definition of 4). But 'if equals be substituted for equals, the equality remains'. Therefore 2 + 2 = 4.

Even if we accept Leibniz's definitions, however, there is a flaw in this proof. It lies in a crucial ambiguity in the expression '2 + 1 + 1'. This can mean either '2 + (1 + 1)', which is what it needs to mean for the step in the proof that depends on the definition of 2 to be correct, or '(2 + 1) + 1', which is what it needs to mean for the step in the proof that depends on the definition of 3 to be correct. It is easy for us to take for granted that there is no difference between these. But consider the following expression involving exponentiation:

> 2 to the power of 2 to the power of 3.

This is similarly ambiguous. It can mean either

> 2 to the power of (2 to the power of 3)

or

(2 to the power of 2) to the power of 3.

In this case the ambiguity matters. There is a significant difference between these two expressions. The first stands for 2^8, which is 256. The second stands for 4^3, which is only 64. We can't simply take for granted that the ambiguity in Leibniz's proof doesn't similarly matter. That is, we can't equate $2 + (1 + 1)$ with $(2 + 1) + 1$ unless we have some justification for doing so. And any such justification must lie precisely in some suitable basic principles and rules for arithmetic. Providing a suitable axiomatization of arithmetic seems to be the only sure way of distinguishing between those arithmetical proofs that really are acceptable and those that merely seem so.

Frege's project

By the mid- to late 19th century, the growing realization that standards of clarity and rigour in mathematics hadn't always been as high as they should have been brought the felt need for such an axiomatization to a head. The most important figure in this connection was Gottlob Frege. Frege's lifetime project was to specify basic principles and rules for the whole of mathematics apart from geometry. (I will say something shortly about why he made an exception of geometry.) This included arithmetic. But the *way* in which it included arithmetic was significant. Frege was not interested in providing an axiomatization of arithmetic as just one among a conglomerate of separate axiomatizations of each of the various non-geometrical branches of mathematics. He was after something more unified than that. He sought a single axiomatization that would capture them all at once.

The reason he made an exception of geometry was the same as the reason he sought to unite everything else. He believed that everything else was really logic in disguise. Geometry deals specifically with space, and therefore makes use of spatial vocabulary. But the whole of the rest of mathematics, Frege believed, made use of nothing but logical vocabulary.

How does this square with the informal characterization of logical vocabulary that I gave in Chapter 2: that it is used in connection with all subjects? Surely it follows from this characterization that *no* branch of mathematics makes use of *nothing but* logical vocabulary? For surely any branch of mathematics has its own distinctive vocabulary, tailored to its own subject? (Arithmetic has '7' and '+', for example.)

There are several things that might be said in response. The informal characterization that I gave was never meant to be anything other than rough. Perhaps Frege's view is correct, and it is the characterization that is at fault. Or perhaps the conflict between them illustrates something to which I have already alluded, that the very distinction between what is logical and what is not is contestable, so that neither Frege's view nor the earlier characterization need be said to be at fault: they just reflect two equally legitimate stances that can be taken on what counts as logic.

Or *perhaps* there is no conflict between them in the first place? This, or something like this, would have been Frege's own response. He would have agreed that logical vocabulary is used in connection with all subjects. But he would have insisted that arithmetical vocabulary is used in connection with all subjects too. Things of all kinds can be counted. And seven things of kind K together with five different things of kind K constitute twelve things of kind K, irrespective of what K is. (Recall that, in Chapter 2, when I introduced a version of the non-Euclidean postulate in which non-logical terms such as 'point' and 'line' had been replaced by alien terms such as 'bumid' and 'ryma', the expression 'more than one' was left intact.) Someone might object, 'What if there were nothing to be counted? Shouldn't logical vocabulary have a use even in talking about nothingness?' Frege's ingenious reply would have been that, even if there were nothing *else* to be counted, there would be the natural numbers themselves: there would be the number 0 registering how many things were not natural numbers; the number 1 registering how many things were not natural numbers other than 0; the number 2 registering how many things were not natural numbers other than 0 or 1; and so on.

But suppose this did not convince. Frege gave additional reasons to support his view. He held that, whatever doubt there might be about whether arithmetical vocabulary such as '7' and '+' counts as logical, the term 'set' certainly does. Things of all kinds can be collected together into sets. (And if someone were to echo the objection above by protesting that there might be nothing to be collected together, a similarly ingenious reply would be forthcoming: even if there were nothing else, there would be the empty set, that's to say the set with no members, together with whatever other sets must exist if it does, such as the set whose only member is the empty set and the set whose only two members are those two.) Frege then argued—the details of his argument don't matter for current purposes—that among the various expressions that can be defined in terms of 'set' and other logical vocabulary there are some which, though they may not initially be recognizable as such, can eventually be identified as items of mathematical vocabulary, including some which can eventually be identified as items of arithmetical vocabulary.

So Frege's project became that of showing that there are basic principles and rules concerning sets, which can serve as basic principles and rules for the whole of non-geometrical mathematics. He supplied what he took to be such basic principles and rules in his masterwork *The Basic Laws of Arithmetic*. Let us call his axiomatization **BLA**. If Frege was right that the vocabulary of **BLA** could be used to define arithmetical vocabulary, then **BLA** furnished precisely the kind of thing to which Gödel's theorem applies: an axiomatization of arithmetic.

The collapse of Frege's project

Given what Gödel went on to establish, we know that either there were arithmetical truths that **BLA** failed to capture or there were arithmetical falsehoods that it did capture. Which? Well, if Frege was *wrong* about whether the vocabulary of **BLA** could be used to define arithmetical vocabulary, then obviously the former. But if he was right about that, then, for reasons that I am about to sketch, the latter. Either way, Frege's project was unsuccessful.

BLA, it turned out, was inconsistent. (Here a point of clarification is called for. An axiomatization is inconsistent, recall, when it is guaranteed to capture some falsehoods purely by virtue of the logical vocabulary involved. So am I now presupposing Frege's view that 'set' counts as logical? No: **BLA** was inconsistent whether he was right about that or not.) The inconsistency was due to a paradox that arose from Frege's (admittedly utterly intuitive) conception of a set. It was Bertrand Russell who discovered the paradox, which is accordingly known as Russell's paradox.

> **Russell's paradox**: A set does not typically belong to itself. For example, the set of mice is not itself a mouse, so it does not belong to itself. On the other hand, some sets do belong to themselves: the set of sets is an example. Now, consider the set of sets of the former kind, those that do not belong to themselves. Call this set R. Does R belong to itself? Well, only if it does *not* belong to itself; for only sets that do not belong to themselves belong to R. So no: R does not belong to itself, on pain of absurdity. But any set that does not belong to itself does belong to R. So yes: R does belong to itself. In sum, R both belongs to itself and does not belong to itself.

Russell communicated this paradox to Frege shortly after the first volume of *Basic Laws of Arithmetic* had been published and while Frege was busy writing the third and final volume. It meant that **BLA** was totally unfit for purpose. It wasn't just that **BLA** could be used to prove a contradiction. The contradiction could not be contained. This was because **BLA** incorporated a formal equivalent of the principle: 'Believe *that* and you'll believe anything'. **BLA** could be used to prove every single statement in its language, including, if Frege was right that its vocabulary could be used to define arithmetical vocabulary, every arithmetical falsehood.

Frege was one of the greatest logicians and mathematicians of all time. (He effectively founded the discipline of logic in its modern guise. Without his pioneering work, what Gödel did would quite literally have been unthinkable.) Even so, the upshot of all of this, and the irony, was that **BLA** served as an instance of the very problem that it had been designed to address: the basic principles and rules that constituted it had seemed totally compelling but had turned out, on closer scrutiny, not to be. Frege died embittered. His life's work seemed to have been a failure.

Principia Mathematica

Meanwhile Russell, who had independently conceived a project very like Frege's, valiantly carried on with it, although of course with a modified conception of a set. (On Russell's conception, the question whether any given set belongs to itself does not even arise, because a set has to be a different *type* of thing from its members.) In collaboration with his former tutor A.N. Whitehead he published *Principia Mathematica*, in which they emulated Frege's attempt to specify basic principles and rules for the whole of non-geometrical mathematics.

This monumental work, which, like *The Basic Laws of Arithmetic*, ran to three volumes, was remarkable for its painstakingness.

After nearly 400 pages of dense prose and esoteric symbolism, there appeared a statement in it labelled *54.43, followed by a proof of the statement, followed by the laconic observation, 'From this [statement] it will follow, when arithmetical addition has been defined, that 1 + 1 = 2'. (See Figure 5.)

Let us call Whitehead and Russell's axiomatization **PM**. Like **BLA**, **PM** furnished precisely the kind of thing to which Gödel's theorem applies. Like **BLA**, it must therefore have either failed to capture some arithmetical truths or captured some arithmetical falsehoods. Unlike **BLA**, however, it has never been shown to be inconsistent. On the assumption that it was *not* inconsistent, and that its vocabulary could be used to define arithmetical vocabulary, its deficiency instead lay in the fact that there were arithmetical statements that it could not be used to prove but whose denials it could not be used to prove either, hence arithmetical truths that, one way or the other, it failed to capture.

It was a variation of **PM** that provided the case study for Gödel's original article. Let us reconsider the title of that article: 'On Formally Undecidable Propositions of *Principia Mathematica* and Related Systems I'. I have already explained the 'I' at the end: we are now in a position to understand the rest of the title. Gödel used the expression 'formal system', or sometimes just 'system', for what I have been calling an axiomatization. And he called any statement in the language of a given system A that A cannot be used to prove or disprove a 'formally undecidable proposition' of A. In his own terminology, then, his article showed that there are formally undecidable propositions of **PM**—the system of *Principia Mathematica*—and of relevantly similar systems.

Hilbert's programme

Another very important figure in the historical background to Gödel's theorem was the mathematician David Hilbert. Hilbert,

*54·42. $\vdash :: \alpha \epsilon 2 . \supset :. \beta \mathbf{C} \alpha . \mathbf{H} ! \beta . \beta \neq \alpha . \equiv . \beta \epsilon \iota``\alpha$

Dem.

$\vdash . *54·4 . \quad \supset \vdash :: \alpha = \iota`x \cup \iota`y . \supset :.$

$\beta \mathbf{C} \alpha . \mathbf{H} ! \beta . \equiv : \beta = \Lambda . \mathbf{v} . \beta = \iota`x . \mathbf{v} . \beta = \iota`y . \mathbf{v} . \beta = \alpha : \mathbf{H} ! \beta :$

[*24·53·56,*51·161] $\qquad \equiv : \beta = \iota`x . \mathbf{v} . \beta = \iota`y . \mathbf{v} . \beta = \alpha$ (1)

$\vdash . *54·25 . \text{Transp} . *52·22 . \supset \vdash : x \neq y . \supset . \iota`x \cup \iota`y \neq \iota`x . \iota`x \cup \iota`y \neq \iota`y :$

[*13·12] $\qquad \supset \vdash : \alpha = \iota`x \cup \iota`y . x \neq y . \supset . \alpha \neq \iota`x . \alpha \neq \iota`y$ (2)

$\vdash . (1) . (2) . \supset \vdash :: \alpha = \iota`x \cup \iota`y . x \neq y . \supset :.$

$\beta \mathbf{C} \alpha . \mathbf{H} ! \beta . \beta \neq \alpha . \equiv : \beta = \iota`x . \mathbf{v} . \beta = \iota`y :$

[*51·235] $\qquad \equiv : (\mathbf{H}z) . z \epsilon \alpha . \beta = \iota`z :$

[*37·6] $\qquad \equiv : \beta \epsilon \iota``\alpha$ (3)

$\vdash . (3) . *11·11·35 . *54·101 . \supset \vdash . \text{Prop}$

*54·43. $\vdash :. \alpha, \beta \epsilon 1 . \supset : \alpha \cap \beta = \Lambda . \equiv . \alpha \cup \beta \epsilon 2$

Dem.

$\vdash . *54·26 . \supset \vdash :. \alpha = \iota`x . \beta = \iota`y . \supset : \alpha \cup \beta \epsilon 2 . \equiv . x \neq y .$

[*51·231] $\qquad \equiv . \iota`x \cap \iota`y = \Lambda .$

[*13·12] $\qquad \equiv . \alpha \cap \beta = \Lambda$ (1)

$\vdash . (1) . *11·11·35 . \supset$

$\vdash :. (\mathbf{H}x, y) . \alpha = \iota`x . \beta = \iota`y . \supset : \alpha \cup \beta \epsilon 2 . \equiv . \alpha \cap \beta = \Lambda$ (2)

$\vdash . (2) . *11·54 . *52·1 . \supset \vdash . \text{Prop}$

From this proposition it will follow, when arithmetical addition has been defined, that $1 + 1 = 2$.

*54·44. $\vdash :. z, w \epsilon \iota`x \cup \iota`y . \supset_{z,w} . \phi(z, w) : \equiv . \phi(x, x) . \phi(x, y) . \phi(y, x) . \phi(y, y)$

Dem.

$\vdash . *51·234 . *11·62 . \supset \vdash :. z, w \epsilon \iota`x \cup \iota`y . \supset_{z,w} . \phi(z, w) : \equiv :$

$z \epsilon \iota`x \cup \iota`y . \supset_z . \phi(z, x) . \phi(z, y) :$

[*51·234,*10·29] $\equiv \phi(x, x) . \phi(x, y) . \phi(y, x) . \phi(y, y) :. \supset \vdash . \text{Prop}$

*54·441. $\vdash :. z, w \epsilon \iota`x \cup \iota`y . z \neq w . \supset_{z,w} . \phi(z, w) : \equiv :. x = y : \mathbf{v} : \phi(x, y) . \phi(y, x)$

Dem.

$\vdash . *5·6 . \supset \vdash :: z, w \epsilon \iota`x \cup \iota`y . z \neq w . \supset_{z,w} . \phi(z, w) : \equiv :.$

$z, w \epsilon \iota`x \cup \iota`y . \supset_{z,w} : z = w . \mathbf{v} . \phi(z, w) :.$

[*54·44] $\qquad \equiv : x = x . \mathbf{v} . \phi(x, x) : x = y . \mathbf{v} . \phi(x, y) :$

$y = x . \mathbf{v} . \phi(y, x) : y = y . \mathbf{v} . \phi(y, y) :$

[*13·15] $\qquad \equiv : x = y . \mathbf{v} . \phi(x, y) : y = x . \mathbf{v} . \phi(y, x) :$

[*13·16,*4·41] $\equiv : x = y . \mathbf{v} . \phi(x, y) . \phi(y, x)$

This proposition is used in *163·42, in the theory of relations of mutually exclusive relations.

5. **The page from *Principia Mathematica* on which Whitehead and Russell refer to their proof that $1 + 1 = 2$.**

like Frege, Whitehead, and Russell, was interested in the project of providing an axiomatization of arithmetic, as part of an axiomatization of a broader swathe of mathematics. He didn't share the aspiration to demonstrate that such mathematics was logic in disguise, but he did share the aspiration to provide it with a visibly secure foundation which would allay any worries mathematicians might feel about once again being led astray by what was speciously compelling—as Frege himself had been.

Hilbert made important contributions of his own towards the attempt to provide such an axiomatization. His chief significance, however, lay in what he had to say *about* such an attempt: about what it would take for such an attempt to succeed and about how we could assess whether it had done so. Unlike his predecessors, he was not ready to say that the success of such an attempt would depend on the truth of all the statements captured by the axiomatization. This was not because he thought that such an attempt could succeed even if the axiomatization captured some falsehoods. It was because he thought that there was a fundamental issue that needed to be addressed about what it is for a mathematical statement to be a candidate for truth or falsity in the first place. Not that his predecessors had been completely insouciant about this issue. Still less had they been completely unselfconscious about it. On the contrary, they had thought very deeply about it. Nevertheless, they had been prepared, in a relatively unregenerate way, to ascribe truth to all the statements captured by their axiomatizations—or, in Frege's case, once Russell's paradox had been discovered, to ascribe truth to some of them and falsity to all the others. But Hilbert thought that a careful critique of what it was for a statement to be a candidate for truth or falsity would not only lead to a reassessment of whether this was as straightforward or as appropriate as it appeared but would also help to show exactly what was required of an axiomatization.

In particular, Hilbert thought that we should not simply take for granted that statements that involved more or less explicit reference to the infinite were the candidates for truth or falsity they appeared to be. The most serious problems that had afflicted earlier mathematics had all directly or indirectly involved the infinite. This includes Russell's paradox. The set R in Russell's paradox, if there were such a set, would have to be infinite, since there are infinitely many sets that do not belong to themselves (such as the set whose only member is 0, the set whose only two members are 0 and 1, the set whose only three members are 0, 1, and 2, and so on). Hilbert, wary of these problems, urged a reconsideration of whether statements about infinite sets even made sense. If they did not, and if they were therefore not to be classified either as true or as false, then we didn't need to worry about the paradoxes that afflicted our attempts to decide which were which.

In this vein, Hilbert began by considering those mathematical statements that he thought *could* unproblematically be classified as true or false—what he called 'finitary' statements. Paradigmatic instances were such elementary arithmetical equations as '7 + 5 = 12', in which specific natural numbers were said to stand in certain relations to one another, and (finite) combinations of these, such as '7 + 5 = 12 and 7 + 5 ≠ 13'. In explaining why he thought that these could unproblematically be classified as true or false, Hilbert gave a very distinctive account of their meaning. In opposition to Frege, Whitehead, and Russell, in whose collective view they were really statements about sets, Hilbert denied that they were statements about anything independent of the language of mathematics. On Hilbert's view, finitary statements had as their subject matter a certain kind of *symbolism*: they reported the results of concatenating and manipulating sequences of signs in various ways. To clarify this, Hilbert envisaged a crude system of numerals whereby each positive natural number was represented by that many strokes. When elementary equations were expressed in this symbolism, they *were* their own meaning. Thus instead of '7 + 5 = 12', we could write

$$| \ | \ | \ | \ | \ | \ | + | \ | \ | \ | \ | = | \ | \ | \ | \ | \ | \ | \ | \ | \ | \ | \ |,$$

and then see by inspection that it was true.

But Hilbert was of course aware that mathematicians made critical use of statements that involved more or less explicit reference to the infinite as well, for instance when making certain general claims about all of the (infinitely many) natural numbers. Hilbert called these 'ideal' statements. What account did he give of such statements? What did he think they meant?

Seizing on the idea mooted above, Hilbert argued that they did not mean anything. They were neither true nor false. They were supplementary devices of a purely formal kind, designed to facilitate proofs and to make for greater elegance and perspicuity. Thus instead of proving in laborious detail the true finitary statement that every natural number less than a million is the sum of four squares, we could take a short cut via the ideal result 'Every natural number is the sum of four squares'.

But there had to be more to it than that. The use of ideal statements could not be justified simply on the grounds that they enabled us to prove true finitary statements in this way. For they could fulfil this function and still be exceptionable—for instance if the only reason why they enabled us to prove true finitary statements was that they enabled us to prove *all* finitary statements, including all finitary falsehoods. Even if ideal mathematics did not itself have to be true, the finitary mathematics that issued from it did.

This gave rise to what became known as Hilbert's programme: to devise an axiomatization **H**, akin to **BLA** and **PM**, such that (i) **H** was powerful enough to enable us to prove all finitary truths but (ii) **H** was *not* powerful enough to enable us to prove any finitary falsehoods. The challenge was to guarantee (i) in such a way as simultaneously to be able to guarantee (ii). And this challenge was exacerbated by the fact that (ii) could not be guaranteed by ensuring that the basic principles in **H** were all true, since some of those basic principles would *not* be true: they would themselves be ideal statements that were neither true nor false. How then could the challenge be met?

By ensuring the consistency of **H**. For suppose that (i) held, but not (ii). Suppose, in other words, that **H** enabled us to prove every finitary truth but also at least one finitary falsehood, say '7 is greater than 12'. Then among the finitary truths that **H** enabled us to prove would be the denial of that falsehood: '7 is not greater than 12'. But this means that **H** would be inconsistent.

A crucial part of Hilbert's programme was therefore to demonstrate **H**'s consistency. Now this calls to mind the discussion at the end of Chapter 2, about how a demonstration of the consistency of a mathematical axiomatization is itself a mathematical exercise. This, we noted, suggests a problematical regress. But, as we also noted, we might at least be able to demonstrate the consistency of one axiomatization by using an axiomatization whose consistency is less open to doubt. This is what Hilbert thought was called for in the current context. What we needed, he argued, was a proof of **H**'s consistency that used basic principles and rules which, unlike those in **H**, were exclusively finitary.

How might this proceed? Not in the same way as the proof of the consistency of the non-Euclidean axiomatization **NA** sketched in Chapter 2. That proof involved specifying a faithful interpretation of **NA**, where by a *faithful interpretation* of an axiomatization I mean a (possibly unintended) interpretation of its non-logical vocabulary that renders it sound. (The interpretation of **NA** in question involved the region above *h* in Figure 4.) The same strategy would not work here. This is because, in order for **H** to fulfil its function, it would have to exploit its logical vocabulary in such a way as to force any faithful interpretation of it to involve the infinite. So we could not specify such an interpretation while restricting ourselves to the use of basic principles and rules that were finitary. The required proof of **H**'s consistency would have to proceed in some other way.

Perhaps such a way could be extracted from Hilbert's general approach to mathematics? Perhaps we could exploit the purely symbolic features of **H**? Here is an unrealistically simple but helpful model. Suppose all the basic principles in **H** had an odd number of symbols. And suppose all **H**'s rules could be shown, by finitary means, to preserve this feature—so that **H** could be used to prove only statements with an odd number of symbols. Finally, suppose we could also see by finitary means that, unless **H** were consistent, it would enable us to prove every statement in its language (because **H** incorporated some formal equivalent of the principle: 'Believe that and you'll believe anything'). Then we would be in a position to construct a finitary proof of **H**'s consistency the moment we found a statement in its language with an even number of symbols.

This was the kind of thing that Hilbert envisaged. And, as we'll see, the various ideas at work here—in particular, the idea that the purely symbolic features of an axiomatization are themselves amenable to mathematical investigation, possibly even to arithmetical investigation—informed Gödel's own work. This relates back to a point I made earlier: although I myself have so far been taking for granted (contra Hilbert) that all mathematical statements can unproblematically be classified as true or false, Gödel's theorem can be stated and proved in a way that bypasses the concept of truth altogether. All of this may appear as justification of sorts for Hilbert's programme. In fact, however, as we'll also see, Gödel's theorem makes serious trouble for the programme.

Chapter 4
The key concepts involved in Gödel's theorem

In this chapter, I will define and explain the key concepts involved in proving Gödel's theorem. (Less precise and less formal versions of some of these concepts have already appeared in earlier chapters; more precise and more formal versions of some of them will appear in later chapters.)

Formal languages

In Chapter 2, we noted that the first thing required of a mathematical axiomatization, from a modern point of view, is a specification of the linguistic resources on which it will draw. We need to make clear exactly what symbolism the axiomatization involves and, more particularly and crucially, which statements are candidates for being proved by means of the axiomatization. Given any axiomatization A, there must be a corresponding language $\mathscr{L}$, such that it is only with respect to statements in $\mathscr{L}$ that there is an issue whether or not they can be proved using A.

Such a language will be what is known as a *formal language*. (I will sometimes omit the word 'formal', where there is no danger of confusion.) A formal language $\mathscr{L}$ has two components:

- a finite lexicon;
- a grammar.

$\mathscr{L}$'s lexicon comprises all its primitive symbols: for example, among the primitive symbols in a language for arithmetic will typically be '0' standing for the natural number 0, '+' standing for addition, and '=' standing for equality. $\mathscr{L}$'s grammar comprises rules for how these primitive symbols can be combined to form statements in $\mathscr{L}$: for example, among the statements in a language for arithmetic will typically be '0 = 0', but not '+ = 0'.

Formal languages differ from natural languages in a number of ways. For instance, a formal language's grammar must *never* leave *any* doubt about whether something is a statement in the language or not. This contrasts with English. Although some strings of words in English are straightforwardly grammatical, like 'Grass is green', and some are straightforwardly ungrammatical, like 'Green are grass', some are borderline—to the extent that there is a roughly even split between those native speakers who count them as *bona fide* statements and those native speakers who don't. Here are two examples: 'His poetry is easy to disparage without even reading'; 'He may have been sleeping, but she may have done so too'.

A further difference between formal languages and natural languages is that, in a natural language like English, there are statements that, though licensed by the grammar, nevertheless lack any significance, for instance 'This parallelogram has toothache', or 'It's midnight at the North Pole'. In a formal language this kind of thing must never happen. Thus a formal language for arithmetic had better not include the symbol '÷', lest there be statements in it, such as '1 ÷ 0 = 0', that have no significance. (Or rather, it had better not include the symbol '÷' unless its grammar is contorted so as to disallow combinations such as this. But that would have disadvantages of its own. The more contorted the grammar, the harder it is to generalize about statements in the language, which is something that in due course we shall need to do.) Here you might object that we can't just exclude '÷' from all formal languages for arithmetic, since it is straightforwardly true that $6 ÷ 3 = 2$ and an axiomatization of arithmetic should capture such truths. This won't be a problem, however, provided we can phrase the truths in question differently, using symbols that are not hostage to the same fortune. In this case, there won't be a problem if the language gives us the wherewithal to say that 2 is the unique natural number n such that $3 × n = 6$. (The symbol '×' is not hostage to the same fortune, since any two natural numbers have a product.)

Note that, although only formal languages are strictly speaking fit to be used in connection with a mathematical axiomatization, *in practice* mathematicians do what I have mostly done when giving examples of arithmetical statements, that's to say they use a mixture of symbols from a formal language and vocabulary appropriated informally from natural language. Also, once the lexicon and grammar of a formal language have been specified, there is nothing to stop mathematicians from making use of certain helpful abbreviations, say calling a natural number 'even' rather than spelling out that it is the result of multiplying some natural number by 2, then treating the language as though it included these abbreviations even though, strictly speaking, it doesn't.

Logical and non-logical vocabulary

Let us reconsider the distinction between logical and non-logical vocabulary from Chapter 2. This distinction applies to the lexicon of any formal language. At various points in what follows I will exploit this fact. Do we therefore have to revisit the philosophical issues about the drawing of the distinction that we considered earlier? No. It doesn't matter for current purposes what *determines* whether any given symbol counts as logical. All that matters is that, given any formal language, some of the symbols in its lexicon count as logical and some do not.

For the record, I will be presupposing a minimalist conception of what counts as logical vocabulary, a conception that nobody seriously contends extends too far but that some people seriously contend doesn't extend far enough. In particular, I will be presupposing (contra Frege) that neither the vocabulary used to talk about sets nor any distinctively arithmetical vocabulary counts as logical. Although this will affect what I have to say, again it doesn't really matter. For it won't make any difference to the *substance* of any of my claims, only to how they are couched. Claims that I shall be making about what I classify as one particular non-logical enterprise—arithmetic—will instead, on certain rival conceptions, need to be recast as claims about the extended reaches of logic itself.

Now there are two symbols that I'm going to stipulate must appear in the logical vocabulary of any formal language and to which there will be appeal in what follows: the symbol ¬ representing negation and the symbol → representing what is called material implication (both of which count as logical even on my minimalist conception). The first is placed before a statement s to form a new statement ¬ s. The second is placed between two statements s and t, with the whole then enclosed in square brackets, to form a new statement $[s → t]$. These symbols are to be understood as follows:

- ¬ s is true if s is false, and it is false if s is true;
- $[s → t]$ is false if s is true and t is false, and it is true otherwise.

So ¬ is a formal equivalent of 'It is not the case that': it converts a truth into a falsehood, and a falsehood into a truth. And → does work that is at least somewhat similar to that done by the construction 'If..., then...'. In particular, when the first three dots in 'If..., then...' are replaced by a truth and the second by a falsehood, the statement as a whole can't be true. There is much controversy concerning how much more than this → and the construction 'If..., then...' have in common. It doesn't matter for our purposes, although it certainly does no great harm, at least as a heuristic, to understand $[s → t]$ as the statement that, if s is true, then t is true.

Key concepts

Logical consequence: Given a formal language $\mathscr{L}$, given a statement s in $\mathscr{L}$, and given a set A of statements in $\mathscr{L}$, s will be said to be a logical consequence of A when the logical vocabulary of $\mathscr{L}$ guarantees that if all the members of A are true, then s is true as well.

Here is an example. Let t and u be two statements in $\mathscr{L}$, and let A contain $[t \to u]$ and $\neg u$. Then $\neg t$ is a logical consequence of A. The sheer meaning of $\to$ and $\neg$ guarantees that if both the members of A are true, then $\neg t$ is true too.

Now the following is a very important result:

The Derivation Theorem: There is a sound finite stock of basic principles and rules that enable us, whenever a given statement s is a logical consequence of a given set of statements A, to establish this and hence to derive s from A.

I won't give the details here. (Actually, there are many different, equally effective candidates, some of them quite unlike one another.) Still less will I prove the result. But some such stock of basic principles and rules—call it a *derivation procedure*—will be taken as given from now on, as will its inclusion in any basic principles and rules that may be proposed for any specific subject. I will also take for granted that, when it comes to providing an axiomatization of arithmetic, there is no need to supplement this derivation procedure with any peculiarly arithmetical basic rules, only with peculiarly arithmetical basic *principles*. (Thus nothing is to be derived from these principles except what follows as a matter of logic.) Taking these things for granted doesn't beg any pertinent questions because, if anything, they make providing a decent axiomatization of arithmetic easier—and so proving Gödel's theorem harder.

Proving the Derivation Theorem, incidentally, was another of Gödel's achievements. He did this in his doctoral thesis.

Theory: A theory is defined as a set of statements in a formal language that contains all its own logical consequences. Given any theory T, its language will be referred to as $\mathscr{L}(T)$.

So a theory T includes whatever, in $\mathscr{L}(T)$, it is logically committed to. Thus if T contains the two statements $[t \to u]$ and $\neg\, u$, then it must contain $\neg\, t$ too.

Decidability (of a set of statements): A set of statements A in a formal language $\mathscr{L}$ will be said to be decidable when there is an algorithm for telling whether any given statement in $\mathscr{L}$ belongs to A.

For example, given any formal language $\mathscr{L}$, the set of statements in $\mathscr{L}$ with an odd number of symbols is decidable: a computer could be programmed to determine whether any given statement belongs to that set. (See Figure 6.) By contrast, the set of *true* statements in $\mathscr{L}$ may well not be decidable.

Note that, when I say that there is an algorithm for telling whether any given statement belongs to a set, I imply no more than that there is an algorithm that can *in principle* be put to this use. *In practice* it may be out of the question to tell whether some given statement has an odd number of symbols—for instance, if the number of symbols in the statement is greater than the number of nanoseconds between now and the heat death of the universe. Part of the force of Gödel's theorem is that it holds even when such issues of practicability are set to one side.

Axiomatizabilty: A theory T will be said to be axiomatizable when there is some decidable set of

statements A in $\mathscr{L}(T)$ such that T consists of the logical
consequences of A. Given any axiomatizable theory T, and
given any such set A, the statements in A can serve as
axioms for T.

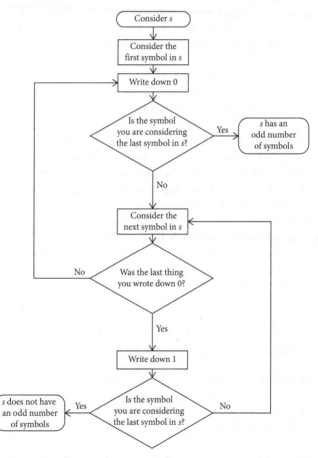

6. **A flowchart to determine whether a given statement s has an odd
number of symbols.**

Notice that this is the first occurrence of the word 'axioms' in this book. This is extremely important. Up until now I have been talking about basic principles. Axioms are very closely related to basic principles. But they are not the same. Nor do they work in the same way.

There are two crucial differences between them. These can best be appreciated in terms of an example. Consider the following:

> Given any property P, if 0 possesses P, and if no natural number possesses P without its successor possessing P as well, then every natural number possesses P.

This might well be regarded as a basic principle of arithmetic. It doesn't follow, however, that any adequate language for arithmetic must include provision for talk about properties. It may not. The point is that, even if it does not, it will still include something corresponding to this basic principle: a decidable set of statements, each of which instantiates the principle with respect to some *specific* property expressible in the language, and each of which can serve as an *axiom* for arithmetic. Here is one such instance (couched here not in its pristine form, but in a mixture of formal vocabulary and English):

> If 0 is less than 2^0, and if there is no natural number n such that n is less than 2^n but its successor $n´$ is not less than $2^{n´}$, then every natural number n is less than 2^n.

There will be infinitely many instances of this kind. This is because there will be infinitely many properties which, like the property of being a natural number n that is less than 2^n, is expressible in the language.

This highlights the two crucial differences between basic principles and axioms:

(1) The basic principles involved in an axiomatization need not themselves be statements in the relevant language, but the axioms involved in it must be.

(2) An axiomatization can't involve more than finitely many basic principles, but it can involve infinitely many axioms.

An axiomatization can involve infinitely many axioms because, although the set of axioms has to be decidable, decidability doesn't entail finitude.

Decidability doesn't *entail* finitude. It nevertheless involves finitude—in as much as, when a set of statements is decidable, there is a finite account of what it takes for a statement to be a member of the set; or, equivalently, there is a finite description of the relevant algorithm. And indeed this is why a decidable set of axioms always *corresponds* to a single basic principle or to a finite set of basic principles, as illustrated in the example above. (This is the subtlety to which I referred in Chapter 1. It is because there is a corresponding basic principle or finite set of basic principles that the statements in a decidable set are fit to serve as axioms at all.)

Something else illustrated in the example above is the idea of a property that is expressible in a language for arithmetic. This leads to the next definition.

> **Arithmetical predicate**: An arithmetical predicate is defined as an expression in a formal language that expresses a property of natural numbers.

To amplify. An arithmetical predicate Π in a formal language $\mathscr{L}$ must contain, as well as symbols from $\mathscr{L}$, one or more occurrences of the ellipsis '_'. When each of these occurrences of '_' is replaced by an expression e in $\mathscr{L}$ that stands for a natural number, what results is a statement s in $\mathscr{L}$ that has e as its grammatical subject. (Note that if there is more than one occurrence of '_' in Π, then each of them is to be replaced by the *same* expression: the resultant statement s will then have a grammatical subject that occurs more than once in it.) Π can be thought of as picking out some property that any given natural number may or may not possess. It applies to any natural number that does possess the property; it fails to apply to any that does not. If it applies to the natural number that e stands for, then the statement s that results when each occurrence of '_' in Π is replaced by e is true; if it fails to do so, then s is false.

Examples of arithmetical predicates (again, couched here not in their pristine form, but in a mixture of formal vocabulary and English, together with some familiar abbreviations) are:

- _ is prime;
- $10 < \_ < 15$;
- _ is the sum of four squares;
- $\_ < 2^-$
  (this was the example considered above);
- _ is odd and _ is divisible by 4;
- $7 + \_ = 12$.

  Characterization: An arithmetical predicate will be said to characterize the set of natural numbers to which it applies.

Thus the six arithmetical predicates above characterize, respectively:

- the set of primes;
- the set containing 11, 12, 13, and 14;
- the set containing every natural number;
- the set containing every natural number
  (this illustrates that a single set can be characterized by more than one arithmetical predicate);
- the set containing no natural numbers whatsoever;
- the set containing just 5.

It is important to note that, given any language $\mathscr{L}$ for arithmetic, although each arithmetical predicate in $\mathscr{L}$ characterizes some set of natural numbers, we can't conclude that each set of natural numbers is characterized by some arithmetical predicate in $\mathscr{L}$—no matter how rich $\mathscr{L}$ is. Some sets may be too 'untidy' to be characterized using $\mathscr{L}$'s resources.

> **Canonical numeral**: Each natural number n has a canonical numeral that stands for it: the symbol **0** preceded by n occurrences of the symbol **S**.

So the canonical numeral standing for 0 is **0**, that standing for 1 is **S0**, that standing for 2 is **SS0**, and so on. We can think of **0** as a simple name for 0, and we can think of **S** as a formal equivalent of 'the successor of'.

There is nothing sacrosanct about this system of numerals: it is simply convenient to have some standard way of referring to individual natural numbers. I could just as well have chosen our own familiar decimal notation, or the binary notation, or a system using '0', '+', and '1', in which the natural numbers are referred to by '0', '(0 + 1)', '((0 + 1) + 1)', and so on.

> **Instance**: Given any arithmetical predicate Π, the statement $\Pi(n)$ that results when each occurrence of the ellipsis in Π is replaced by the canonical numeral that stands for n will be said to be an instance of Π.

Examples of instances of the six arithmetical predicates cited above are:

- **SSS0** is prime. (This is true.)
- $10 <$ **SSSSSSS0** < 15. (This is false.)
- **0** is the sum of four squares. (This is true.)
- **S0** $< 2^{S0}$. (This is true.)

- **SSSSSSSS0** is odd and **SSSSSSSS0** is divisible by 4. (This is false.)
- 7 + **SSSSS0** = 12. (This is true.)

> **Decidability (of a set of natural numbers)**: A set of natural numbers E will be said to be decidable when there is an algorithm for telling whether any given natural number belongs to E.

The set of even numbers, the set of primes, and the set containing just 11, 12, 13, and 14 are examples of decidable sets of natural numbers. (See Figure 7.)

> **Sufficient richness**: A formal language will be said to be sufficiently rich when it contains: (i) all the canonical numerals; and (ii) for each decidable set of natural numbers E, an arithmetical predicate that characterizes E.

Sufficient richness is a property of formal languages. In effect it is the property of containing some basic arithmetical vocabulary. There is a closely related property of *theories* which I will define next and which is an embellishment of it: sufficient strength. For a theory to be sufficiently strong is not just for its language to be sufficiently rich, but also for the theory itself to contain certain core truths belonging to the language. Sufficient strength plays the more significant rôle in what follows, and I could in fact have proceeded straight to the definition of that. But I have included the definition of sufficient richness too because, as we'll see, there are ways of couching and proving Gödel's theorem for which it suffices.

> **Sufficient strength**: A theory T will be said to be sufficiently strong when $\mathcal{L}(T)$ contains: (i) all the canonical numerals; and (ii) for each decidable set of natural numbers E, an arithmetical predicate Π that characterizes E *and* that satisfies the following condition: given any natural number n, if $\Pi(n)$ is true, then T contains $\Pi(n)$, and if $\Pi(n)$ is false, then T contains $\neg\ \Pi(n)$.

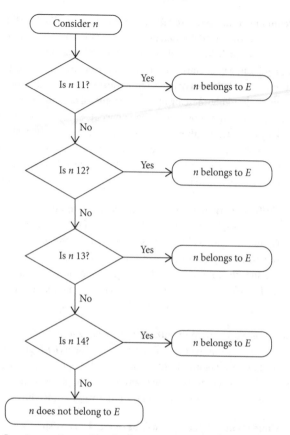

7. **A flowchart to determine whether a given natural number _n_ belongs to the set _E_ containing just 11, 12, 13, and 14. (This flowchart, though crude, testifies to the fact that any finite set of natural numbers is decidable.)**

It is important to appreciate that, although my definition of sufficient strength makes use of the concept of truth, there is a modified definition of it that does not. This is important in connection with my repeated claim that Gödel's theorem can be stated and proved in a way that bypasses the concept of truth. Actually, it is not just my definition of sufficient strength that needs to be modified for this purpose. So do: my definition of characterization, in which 'applying to' amounts to 'being *true* of'; my definition of the canonical numerals, in which 'standing for' amounts to something similar; and of course my definition of logical consequence. Due modifications are available. Roughly, these involve renouncing all semantic relations, that's to say all relations that depend on what expressions mean, and making do instead with mere correlations between features of expressions and features of natural numbers—or at least they do in all but the case of logical consequence, to which I will return shortly. What this comes to in the case of sufficient strength is the following.

> **Sufficient strength (non-semantically defined)**: A theory T can be said to be sufficiently strong when $\mathscr{L}(T)$ contains: (i) all the canonical numerals; and (ii) for each decidable set of natural numbers E, an arithmetical predicate Π that satisfies the following condition: given any natural number n, if n belongs to E, then T contains $\Pi(n)$, and if n does not belong to E, then T contains $\neg\,\Pi(n)$.

Here there is no mention either of characterization or of truth. All that matters is that whether T contains $\Pi(n)$ or $\neg\,\Pi(n)$ is correlated with whether n belongs to E or not. As for logical consequence, we can exploit the Derivation Theorem: we can construe logical consequence as whatever the relevant derivation procedure enacts.

> **Consistency**: A theory T will be said to be consistent when, given any statement s in $\mathscr{L}(T)$, at most one of s and $\neg\,s$ belongs to T.

> **Completeness**: A theory T will be said to be complete when, given any statement s in $\mathscr{L}(T)$, at least one of s and $\neg\,s$ belongs to T.

> **Inconsistency**: A theory T will be said to be inconsistent when it is not consistent, that is when there is some statement s in $\mathscr{L}(T)$ such that both s and $\neg\,s$ belong to T.

Incompleteness: A theory T will be said to be incomplete when it is not complete, that is when there is some statement s in $\mathscr{L}(T)$ such that neither s nor $\neg\, s$ belongs to T.

In Chapter 2, consistency and inconsistency were reckoned to be properties of axiomatizations, not of theories. They were also understood in terms of logical vocabulary in general, not of $\neg$ in particular. We could, if we wanted, think of the definitions above as simply introducing two new homonymous concepts. That said, there are clear and obvious connections between the concepts introduced in Chapter 2 and the concepts defined above; and the latter do all the important work that was being done by the former, though more crisply and more formally.

Notice that the definition of consistency allows for the possibility that there is a statement s in $\mathscr{L}(T)$ such that neither s nor $\neg\, s$ belongs to T, in other words it allows for the possibility that T is incomplete, while the definition of completeness allows for the possibility that there is a statement s in $\mathscr{L}(T)$ such that both s and $\neg\, s$ belong to T, in other words it allows for the possibility that T is inconsistent. This is significant. It signals that, while consistency and completeness may both be regarded as desiderata of any theory, they somewhat militate against each other and each may be achieved at the expense of the other.

A new statement of Gödel's theorem

Granted these definitions, Gödel's theorem can be restated as follows:

> No theory can be sufficiently strong, consistent, complete, and axiomatizable.

The connections with my initial statement of the theorem in Chapter 1 are clear: being sufficiently strong means containing some core arithmetical truths; being consistent is a precondition of containing nothing but the truth; and being complete is a precondition of containing the whole truth (in the theory's language).

One point that needs to be flagged is that, despite the connections just outlined, this new statement of the theorem doesn't directly involve the concept of truth at all. (Recall the note above on the definition of sufficient strength.) Admittedly, neither the concept of consistency nor the concept of completeness would have the significance it has if ¬ didn't mean what it does, which is a matter of truth. Even so, this statement of the theorem doesn't itself depend on what ¬ means; nor does its proof.

We can achieve a better sense of what is at stake in this statement of Gödel's theorem by considering each of the four conditions mentioned in it and finding an example of a theory that satisfies all but that one.

All but sufficient strength: Both analysis and geometry furnish examples, as I mentioned in Chapters 1 and 2, respectively. For that matter, there are examples involving arithmetic. Thus imagine a severely restricted language for arithmetic that contains no statements about any operations other than addition: the set of truths in that language can be shown to satisfy every condition but sufficient strength.

All but consistency: A suitably modified version of Frege's **BLA**, which we considered in Chapter 3, would furnish an example. Indeed, given any sufficiently rich language $\mathscr{L}$, the set containing every statement in $\mathscr{L}$ is an example. (This set is axiomatizable because it can serve as its own set of axioms, and it can do this because it is trivially decidable.)

All but completeness: A suitably modified version of Whitehead and Russell's **PM**, which we considered in Chapter 3, would furnish an example—provided it is

consistent, which we have no reason to suppose it is not.
Another important example will be considered in Chapter 6.

All but axiomatizabilty: Given any sufficiently rich language
$\mathscr{L}$, the set of truths in $\mathscr{L}$ is an example. (This is a reformulation
of my initial statement of Gödel's theorem in Chapter 1.)

Gödel numbering

There is one more concept that I need to define in this chapter,
because it is integral to the proof of Gödel's theorem, if not to its
content.

Gödel numbering: A Gödel numbering is an assignment of
natural numbers to the expressions in a formal language
such that: (i) no two expressions are assigned the same
natural number; and (ii) there is an algorithm for
determining the natural number that is assigned to any
given expression and for determining the expression, if any,
to which any given natural number is assigned.

There are all sorts of ways of constructing Gödel numberings.
One way that is suitable for any formal language containing no
more than ninety primitive symbols is to assign a two-digit
number to each primitive symbol—this assignment can be
quite arbitrary—and then to proceed by means of simple
concatenation. For instance, suppose the following is a statement
in the language:

$$\neg 0 = S0.$$

And suppose $\neg$ is assigned 21, **0** is assigned 43, = is assigned 37,
and **S** is assigned 50. Then the statement as a whole is assigned
2143375043—or, more perspicuously, 2,143,375,043.

Another way of constructing a Gödel numbering is to raise
successive prime numbers to the powers of the natural numbers

assigned to successive primitive symbols in the expression and then to assign the product of the results to the expression as a whole. Given the assignment of natural numbers to primitive symbols considered above, the statement ¬ **0** = **S0** is then assigned the (very large) natural number:

$$2^{21} \times 3^{43} \times 5^{37} \times 7^{50} \times 11^{43}.$$

This method works because, given any natural number n greater than 1, there is a unique way of decomposing n into prime factors, which in turn ensures that there is an algorithm for determining the expression, if any, to which n is assigned.

We'll see in due course the work Gödel numberings can do, but we are already in a position to note one way in which they can be exploited: granted a particular Gödel numbering, the decidability of a set of statements can be thought of as the decidability of its corresponding set of natural numbers. This is significant since we shall eventually need to give a more precise account of decidability and this will save us from having to do so twice.

Note that any Gödel numbering *can also be extended to proofs*. We have left open what form proofs take, but it doesn't matter. A Gödel numbering can be extended to them whatever form they take. The simplest case is that in which a proof is construed as a sequence of statements, each of which is either an axiom or derivable from predecessors in accord with the relevant derivation procedure. The proof can then be assigned a natural number using the prime number method just illustrated, with successive prime numbers raised to the powers of the natural numbers assigned to successive statements in the proof.

Chapter 5
The diagonal proof of Gödel's theorem

This chapter will present one proof of Gödel's theorem—what I will call the diagonal proof—Chapter 6 another. But even this chapter will present two versions of its proof. The first version will proceed on the assumption that any statement in any given formal language is either true or false. The second will dispense with that assumption and proceed in a way that respects my repeated claim that Gödel's theorem can be stated and proved in a way that entirely bypasses the concept of truth and other semantic concepts.

The diagonal proof is much simpler than what we find in the bulk of Gödel's article, which is what I will present in Chapter 6. It nevertheless has its roots in an informal sketch that Gödel gives, in §1 of his article, of the more complicated proof that is to come. The more complicated proof has various advantages over this one: it exploits more rigorous definitions of some of the key concepts; it enables us to apply Gödel's theorem to particular axiomatizations in such a way as to identify statements that these axiomatizations cannot be used either to prove or to disprove; and it can be extended to a proof of the second incompleteness theorem which I mentioned at the end of Chapter 1. Later I will mention a fourth advantage. But none of this should detract from the significance of what follows.

The semantic version of the proof

My original informal statement of Gödel's theorem was that no axiomatization can determine the whole truth and nothing but the truth concerning arithmetic. In Chapter 4, I commented that this can be reformulated as follows: *given any sufficiently rich formal language $\mathscr{L}$, the set of truths in $\mathscr{L}$ is not axiomatizable.* This is what I will prove in this section. (As we'll see, some of the key concepts I defined in Chapter 4 will not be required for this purpose: in particular, there will be no need to use the concept of sufficient strength, or that of (in)consistency, or that of (in)completeness.)

Let $\mathscr{L}$ be a sufficiently rich language. And let **T** be the set of truths in $\mathscr{L}$.

Assume **T** is axiomatizable. (My aim is to derive a contradiction from this assumption, and thereby to conclude that **T** is not axiomatizable.)

Given that **T** is axiomatizable, there must be some decidable set of statements **A** in $\mathscr{L}$ that can serve as a set of axioms for **T**. From this we can infer that **T** is *itself* decidable. That is, there is an algorithm α for telling whether any given statement s in $\mathscr{L}$ belongs to **T**—or, equivalently, for telling whether any given statement s in $\mathscr{L}$ is true. To see why, let **G** be a Gödel numbering for $\mathscr{L}$ and for proofs involving $\mathscr{L}$, and call the natural number that **G** assigns to any given expression or proof its **G**-number. Now, given that **A** can serve as a set of axioms for **T**, and given that either s belongs to **T** or $\neg s$ belongs to **T**, we can conclude that either s is a logical consequence of **A** or $\neg s$ is a logical consequence of **A**. So, given the Derivation Theorem, either there is a proof that s is a logical consequence of **A** or there is a proof that $\neg s$ is a logical consequence of **A**. The algorithm α is to start with 0 and to consider successive natural numbers, ascertaining in each case whether it is the **G**-number of a proof of either of these kinds. Eventually there will come a natural number n that is. If n is the **G**-number of a proof of the former kind, that is a proof that s is a logical consequence of **A**, then s belongs to **T**. If n is the **G**-number of a proof of the latter kind, that is a proof that $\neg\, s$ is a logical consequence of **A**, then $\neg\, s$ belongs to **T**, and hence, since s and $\neg\, s$ cannot both belong to **T**, s does *not* belong to **T**. (See Figure 8.)

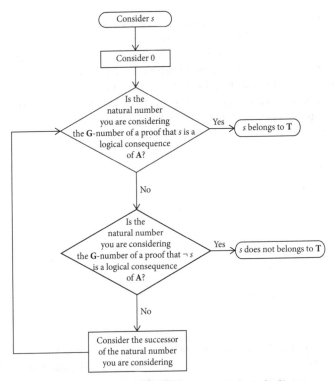

8. A flowchart to determine whether a given statement s in $\mathscr{L}$ belongs to T.

Now consider the following recipe for listing all the (infinitely many) arithmetical predicates in $\mathscr{L}$. Start with 0 and consider successive natural numbers, ascertaining in each case whether it is the **G**-number of an arithmetical predicate in $\mathscr{L}$. If it is, add that arithmetical predicate to the list. So first on the list is the arithmetical predicate with the lowest **G**-number—call it Π_0. Second on the list is the arithmetical predicate with the second lowest **G**-number—call it Π_1. And so on. Given the nature of **G**, there is an algorithm for telling, given any natural number n, which arithmetical predicate is Π_n.

We can now consider an infinite table τ of *yeses* and *noes* registering whether or not successive arithmetical predicates on this list apply to successive natural numbers. Here is an arbitrarily chosen example of what the top left-hand corner of τ might look like. (What τ actually looks like will depend on $\mathscr{L}$ and **G**. In the example given, Π_0 applies to 0, but not to 1, 2, or 3. By contrast, Π_1 applies to 1, 2, and 3, but not to 0. So perhaps Π_0 is some formal version of '_ = 0', while Π_1 is some formal version of '_ ≠ 0'.)

	0	1	2	3	...
Π_0	yes	no	no	no	...
Π_1	no	yes	yes	yes	...
Π_2	no	yes	no	yes	...
Π_3	yes	yes	no	no	...
.	.	.	.	.	...
.	.	.	.	.	...
.	.	.	.	.	...

T's decidability ensures that there is an algorithm for telling whether there is a *yes* or a *no* at any point in the table, that is, for telling whether any given arithmetical predicate applies to any given natural number. Thus suppose we want to tell whether there

66

is a *yes* or a *no* at the point where the row for Π_{821} meets the column for 243. We first determine, algorithmically, which arithmetical predicate is Π_{821}, then determine, using the algorithm α, whether Π_{821} applies to 243, that is whether the statement $\Pi_{821}(243)$ is true. If this statement is true, there is a *yes* at that point; if it is false, there is a *no* there.

Now consider the infinite sequence of *yeses* and *noes* that constitute the diagonal of τ, starting at the top left-hand corner.

	0	1	2	3	...
Π_0	yes	no	no	no	...
Π_1	no	yes	yes	yes	...
Π_2	no	yes	no	yes	...
Π_3	yes	yes	no	no	...
.	.	.	.	.	...
.	.	.	.	.	...
.	.	.	.	.	...

In this example the sequence is as follows:

yes, yes, no, no, ...

And consider the infinite sequence of *yeses* and *noes* in which these are reversed, which in this example is:

no, no, yes, yes, ...

This sequence of *yeses* and *noes*, just like the sequence of *yeses* and *noes* on any row in τ, corresponds to a set of natural numbers—a set which in this example contains 2 and 3 but doesn't contain either 0 or 1. Call this set **D**. Then **D** is decidable. This is because telling whether a given natural number n belongs to **D** is tantamount to telling whether there is a *no* at the point in τ where the row for Π_n meets the column for n—which, as we have just seen, there is an algorithm for doing. For example, we can tell that 2 belongs to **D** because we can tell that there is a *no* at the point where the row for Π_2 meets the column for 2.

But if **D** is decidable, then $\mathscr{L}$'s sufficient richness means that there must be an arithmetical predicate Π_i in $\mathscr{L}$ that characterizes **D**.

This, however, is impossible. By construction, none of the arithmetical predicates on the list—and they are all there—has the right sequence of *yeses* and *noes* on its row. The sequence of *yeses* and *noes* on the row for Π_0, for instance, differs from that which corresponds to **D** in its first place: it has a *yes* there rather than a *no*. The sequence of *yeses* and *noes* on the row for Π_1 differs from that which corresponds to **D** in its second place: again, it has a *yes* there rather than a *no*. The sequence of *yeses* and *noes* on the row for Π_2 differs from that which corresponds to **D** in its third place: it has a *no* there rather than a *yes*. And so on. The original supposition that **T** is axiomatizable must therefore be rejected. *Q.E.D.*

Three noteworthy features of the proof

There are three features of the proof just given that deserve special mention.

(i) *The (un)decidability of* **T**: It should have seemed striking that we could infer the decidability of **T** from its axiomatizability. In general, axiomatizability is one thing, decidability another. The decidability of a theory means that there is a purely mechanical way of telling whether any given statement belongs to the theory. Not so its axiomatizability. Suppose you are presented with a finite stock of basic principles and rules, together with some statement that you are assured can be proved by means of them, and are challenged to verify this. Only due imagination and insight, together perhaps with a little luck, will in general enable you to rise to the challenge—as anyone who has ever struggled with their mathematics homework knows all too well! But in the case of **T**, the peculiarity that it comprises the whole truth and nothing but the truth within its language means that, had it been axiomatizable, it would thereby have been decidable too.

It is tempting to think that this takes some of the sting out of Gödel's theorem. For it is tempting to think that, however reasonable it may have seemed to hope that **T** was axiomatizable, it ought never to have seemed reasonable to hope that **T** was *decidable*. We do not in general acquire knowledge of arithmetical truths algorithmically. Admittedly, we acquire knowledge of *some* arithmetical truths algorithmically. For example, we know that $827 \times 240 = 198{,}480$ because we have calculated this using the algorithm for multiplication. But other arithmetical truths that we know, for example that every natural number is the sum of four squares, we know only because some arithmetician, in this case Lagrange, was inspired and skilled enough to divine the proof. Had there been an algorithm for distinguishing between truth and falsity in arithmetic, this would have put the likes of Lagrange out of a job—no?

No. Nor does anything in these considerations take any of the sting out of Gödel's theorem. Just as the existence of a set of axioms for a theory is in general one thing and the existence of an algorithm for telling whether a statement belongs to the theory another, so too the existence of such an algorithm is in general one thing and our actually implementing it another. The algorithm α, had it existed, would have been hopelessly unwieldy, even for a computer. There would never have been any question *in practice* of our taking advantage of it. It would no more have put arithmeticians out of a job than the actual existence of an algorithm for distinguishing between truth and falsity in other branches of mathematics puts their practitioners out of a job, nor, for that matter, than the actual existence of an algorithm for playing perfect chess—there *is* such a thing, because, provided that draws are always claimed, there are only finitely many possible chess games—deprives chess of its interest or prevents there from being grandmasters who are especially good at it. As I remarked in Chapter 4, part of the force of Gödel's theorem is that it holds *even when* these issues of practicability are set to one side.

(ii) *Telling whether there is a yes or a no at any point in* τ: The temptation to underestimate the shock value of there not being any such algorithm as α is accompanied by the converse temptation to *over*estimate the shock value of there not being an algorithm for telling whether there is a *yes* or a *no* at any point in the table τ. For it is tempting to think that, whatever inspiration and skill were required of Lagrange to prove his result, or whatever inspiration and skill may yet be required to prove, say, Goldbach's conjecture—if Goldbach's conjecture is true—we might reasonably have expected there to be an algorithm for distinguishing between truth and falsity within the restricted range of statements involved in τ, that is statements to the effect that some particular natural number, 13 say, possesses some particular property, being prime say.

But this temptation too must be resisted. If $\mathcal{L}$ is a typical language for arithmetic, then being able to tell whether there is a *yes* or a *no* at any point in τ would itself have entailed being able to tell whether Goldbach's conjecture is true. Why? Because if $\mathcal{L}$ is a typical language for arithmetic, then one of the arithmetical predicates in $\mathcal{L}$ will be some formal version of '_ divides a natural number greater than _ which is not the sum of two primes', and that applies to 2 if and only if Goldbach's conjecture is false. For that matter, although I have never stated with precision what counts as an arithmetical predicate, on any standard way of making this precise some formal version of the following would count: '_ = _ and every even number greater than 2 is the sum of two primes'. That applies to 0 (say) if and only if Goldbach's conjecture is true.

(iii) *The non-characterizability of* **D**: The proof finishes with a demonstration that no arithmetical predicate in $\mathscr{L}$ characterizes **D**, which, on pain of violation of $\mathscr{L}$'s sufficient richness, shows that **D** is undecidable. But at this point a concern is liable to arise (if it didn't arise a while ago). The concern is that the very requirement that $\mathscr{L}$ be sufficiently rich, and in particular that $\mathscr{L}$ contain an arithmetical predicate characterizing each decidable set of natural numbers, is inappropriately demanding; more specifically, that it is too demanding to be reasonably imposed on any formal language for arithmetic. In Chapter 4, I casually suggested that for a formal language to be sufficiently rich is just for it to contain basic arithmetical vocabulary. But why think that every decidable set of natural numbers should be characterizable using 'basic arithmetical vocabulary'? In particular, why think this if the algorithm for telling whether any given natural number belongs to the set seems to be nothing like that for telling whether any given natural number is prime, say, but rather involves appeal to some Gödel numbering and the properties of some theory? If this concern is well-founded, this doesn't prevent the result in the previous section from being of interest. But it does prevent it from being a result *about arithmetic*. It leaves open the possibility that the set of truths in some standard language for arithmetic *is* axiomatizable—the point being that the diagonal proof cannot be adapted to rule this out, since there is no reason why the set of natural numbers obtained by reversing the *yeses* and *noes* on the relevant diagonal should not be both decidable (as the axiomatizability of the set of truths would require it to be) and yet uncharacterizable by any arithmetical predicate in the language.

We can't give an adequate response to this concern without a more precise definition of decidability. That must await Chapter 6, where we shall eventually be in a position to assuage the concern. And indeed that is the other (fourth) advantage that the more complicated proof of Gödel's theorem has over the diagonal proof.

The non-semantic version of the proof

I will now present a version of the diagonal proof that bypasses the concept of truth and other semantic concepts. I will prove what I dubbed in Chapter 4 the 'new statement' of Gödel's theorem, which, as I indicated, is suitably non-semantic: *no theory can be sufficiently strong, consistent, complete, and axiomatizable.*

Assume **T*** is a theory that is sufficiently strong, consistent, complete, and axiomatizable. (As before, my aim is to derive a contradiction from this assumption, and thereby to conclude that there can't be any such theory.)

Just as in the semantic version of the proof, we begin with an argument that, because **T*** is axiomatizable, and because there is therefore some decidable set of statements **A*** in $\mathscr{L}(\mathbf{T^*})$ that can serve as a set of axioms for **T***, **T*** must be decidable. The argument for this is almost identical to its counterpart in the semantic version of the proof. It differs critically at just two points, where it involves an appeal to **T***'s completeness or consistency rather than to the exhaustive or the exclusive truth of **T***'s statements. In what follows I will underline each of these.

Let **G*** be a Gödel numbering for $\mathscr{L}(\mathbf{T^*})$ and for proofs involving $\mathscr{L}(\mathbf{T^*})$, and call the natural number that **G*** assigns to any given expression or proof its **G***-number. And let s be an arbitrary statement in $\mathscr{L}(\mathbf{T^*})$. Now, given that **A*** can serve as a set of axioms for **T***, <u>and given that **T*** is complete</u>, which means that either s belongs to **T*** or $\neg s$ belongs to **T***, we can conclude that either s is a logical consequence of **A*** or $\neg s$ is a logical consequence of **A***. So, given the Derivation Theorem, either there is a proof that s is a logical consequence of **A*** or there is a proof that $\neg s$ is a logical consequence of **A***. The algorithm for telling whether s belongs to **T*** is to start with 0 and to consider successive natural numbers, ascertaining in each case whether it is the **G***-number of a proof of either of these kinds. Eventually there will come a natural number n that is. If n is the $\mathbf{G^*}$-number of a proof of the former kind, that is a proof that s is a logical consequence of $\mathbf{A^*}$, then s belongs to $\mathbf{T^*}$. If n is the $\mathbf{G^*}$-number of a proof of the latter kind, that is a proof that $\neg s$ is a logical consequence of $\mathbf{A^*}$, then $\neg s$ belongs to $\mathbf{T^*}$, and hence, <u>since $\mathbf{T^*}$ is consistent</u>, which means that s and $\neg s$ cannot both belong to $\mathbf{T^*}$, s does *not* belong to $\mathbf{T^*}$.

What we have just seen is indicative. The whole of the non-semantic version of the proof proceeds in the same way as the semantic version *mutatis mutandis* wherever the semantic version involves some appeal to a semantic concept. And in fact there are only two further points at which modification of the semantic version of the proof is required.

First, although we can consider a counterpart τ^* of the infinite table τ, this counterpart τ^* cannot register whether or not successive arithmetical predicates apply to successive natural numbers. For that would depend on what the arithmetical predicates meant: it would be tantamount to registering whether the corresponding instances of those arithmetical predicates were true. However, τ^* can do something close. Truth, in the semantic version of the proof, is the same as membership of $\mathbf{T}$. And τ^* can register whether or not successive instances of successive arithmetical predicates belong to $\mathbf{T^*}$. (So, for example, there is a *yes* at the point in τ^* where the row for $\mathbf{\Pi^*}_n$ meets the column for m if the statement $\mathbf{\Pi^*}_n(m)$ belongs to $\mathbf{T^*}$, and there is a *no* there otherwise.) Membership of $\mathbf{T^*}$, it will turn out, can play the same rôle as was played in the semantic version of the proof by membership of $\mathbf{T}$, even though it is not itself a matter of truth.

Second, although we can construct a set $\mathbf{D^*}$ containing all and only those natural numbers n such that there is a *no* at the point on the diagonal of τ^* where the row for $\mathbf{\Pi^*}_n$ meets the column for n, and although we can infer from $\mathbf{T^*}$'s decidability that $\mathbf{D^*}$ is decidable, we cannot appeal to the sufficient richness of $\mathscr{L}(\mathbf{T^*})$ to infer that $\mathscr{L}(\mathbf{T^*})$ contains an arithmetical predicate $\mathbf{\Pi^*}_i$ that characterizes $\mathbf{D^*}$; for characterization is another semantic concept. But we can once again do something close. We can appeal to $\mathbf{T^*}$'s sufficient strength, understood in the non-semantic way specified in Chapter 4, to infer that $\mathscr{L}(\mathbf{T^*})$ contains an arithmetical predicate $\mathbf{\Pi^*}_i$ that satisfies the following condition: given any natural number j, if j belongs to $\mathbf{D^*}$, then $\mathbf{T^*}$ contains $\mathbf{\Pi^*}_i(j)$, and if j does not belong to $\mathbf{D^*}$, then $\mathbf{T^*}$ contains $\neg \, \mathbf{\Pi^*}_i(j)$, which means, given $\mathbf{T^*}$'s consistency, that $\mathbf{T^*}$ does not contain $\mathbf{\Pi^*}_i(j)$. As before, the rôle played in the semantic version of the proof by truth is now in effect being played by membership of $\mathbf{T^*}$. And, as before, this is a rôle that membership of $\mathbf{T^*}$ can play even though it is not a matter of truth.

In particular, and crucially, there is enough here to generate the relevant contradiction. For no arithmetical predicate $\mathbf{\Pi^*}_i$ in $\mathscr{L}(\mathbf{T^*})$ *can* satisfy the specified condition. By construction, none of them has the right sequence of *yeses* and *noes* on its row in $\mathbf{\tau^*}$. The first fails in its first place; the second fails in its second place; the third fails in its third place; and so on. The original supposition that $\mathbf{T^*}$ is sufficiently strong, consistent, complete, and axiomatizable must therefore be rejected. *Q.E.D.*

Chapter 6
A second proof of Gödel's theorem, and a proof of Gödel's second theorem

The two versions of the diagonal proof in Chapter 5 yielded two slightly different results, one semantic and one non-semantic. The semantic result was that the set of truths in a sufficiently rich formal language is not axiomatizable. The non-semantic result was that no theory can be sufficiently strong, consistent, complete, and axiomatizable.

The main proof in Gödel's article yielded an intermediate result. Gödel showed that no theory can be sufficiently strong, *sound*, complete, and axiomatizable—where for a theory to be sound is for all the statements in it to be true. (The concept of soundness has so far been applied only to axiomatizations, but this is the natural extension of it to theories.) This differed from the non-semantic result by including the concept of soundness in place of that of consistency. Not that Gödel's encroachment into semantic territory was great. For one thing, as we'll see, he captured all that mattered about soundness, at least for his purposes, in non-semantic terms. For another, as we'll also see, the encroachment in any case turned out to be ultimately inessential: Gödel's proof could be adapted to prove the original non-semantic result too.

Gödel's strategy was as follows. He showed how, given a particular theory T that satisfies three of the relevant conditions—sufficient strength, soundness, and axiomatizability—we can construct a statement s in $\mathscr{L}(T)$ such that neither s nor $\neg\, s$ belongs to T, in other words a statement s that shows T not to satisfy the fourth condition, completeness. Gödel took as his sample theory a version of Whitehead and Russell's theory which I introduced in Chapter 3. But the techniques that he used could be readily applied to other theories. In this chapter, I will take as my sample theory a theory that I will call **PA**. **PA** is a variant of Peano Arithmetic, which is named after the mathematician Giuseppe Peano, who presented an axiomatization of it late in the nineteenth century, and which is nowadays regarded as a (if not the) paradigmatic arithmetical theory.

The axiomatizability of **PA**

$\mathscr{L}(\mathbf{PA})$ contains the following five non-logical primitive symbols:

- **0** standing for the natural number 0;
- **S** standing for successorship (it is a condition of **PA**'s sufficient strength that it should contain these first two symbols);
- = standing for equality;
- + standing for addition (this is always accompanied by two rounded brackets indicating its scope);
- × standing for multiplication (this too is always accompanied by two rounded brackets indicating its scope).

$\mathscr{L}(\mathbf{PA})$ also contains the following primitive symbol, which is normally classified as logical since it can appear in any formal language irrespective of subject matter, although its meaning is partially determined by context and, in the context of the five non-logical primitive symbols listed above, it registers that the subject matter of the language is the natural numbers:

∀ meaning 'given any natural number'.

Here is a list of statements that can serve as axioms for **PA**:

- $\forall x\ x = x$. (This says that every natural number equals itself.)
- $\forall x\ \neg\ \mathbf{S}x = \mathbf{0}$. (This says that no natural number has 0 as its successor.)
- $\forall x\ \forall y\ [\neg\ x = y \rightarrow \neg\ \mathbf{S}x = \mathbf{S}y]$. (This says that no two natural numbers have the same successor.)
- $\forall x\ (x + \mathbf{0}) = x$. (This says that the sum of any natural number and 0 is that natural number.)
- $\forall x\ \forall y\ (x + \mathbf{S}y) = \mathbf{S}(x + y)$. (This says that the sum of any natural number and the successor of any natural number is the successor of their sum.)
- $\forall x\ (x \times \mathbf{0}) = \mathbf{0}$. (This says that the product of any natural number and 0 is 0.)
- $\forall x\ \forall y\ (x \times \mathbf{S}y) = ((x \times y) + x)$. (This says that the product of any natural number and the successor of any natural number is the sum of their product and the first of them.)

This is not yet the complete list. Recall the discussion in Chapter 4 of the distinction between basic principles and axioms. There are infinitely many further statements on the list corresponding to each of the two following basic principles:

(1) Given any property P, if a natural number possesses P, then so does whatever is equal to it. (For instance, if 12 is even, and if $7 + 5 = 12$, then $7 + 5$ is even.)

(2) Given any property P, if 0 possesses P, and if no natural number possesses P without its successor possessing P as well, then every natural number possesses P. (This is the basic principle that I used to illustrate the distinction between basic principles and axioms in Chapter 4.)

Before I specify the additional axioms more precisely, I need to introduce a notational convention. Given any arithmetical predicate Π in $\mathscr{L}(\mathbf{PA})$, and given any expression e in $\mathscr{L}(\mathbf{PA})$ that can combine with Π as its grammatical subject, including any variable, let 'Π/e' stand for what results when each occurrence of

77

'_' in Π is replaced by e. ('Π/e' can be read as 'Π applies to e'.) For instance, if Π is

$$(\_ \times \mathbf{0}) = \mathbf{0},$$

and if e is

SSS0,

then Π/e is

$$(\mathbf{SSS0} \times \mathbf{0}) = \mathbf{0}.$$

(Notice that '$\Pi/\mathbf{SSS0}$' and '$\Pi(3)$' are two ways of referring to the same statement.) Again, if Π is

$$(\_ + \mathbf{SS0}) = \mathbf{SSSS0},$$

and if e is

x,

then Π/e is

$$(x + \mathbf{SS0}) = \mathbf{SSSS0}.$$

The additional axioms can now be specified as follows. Given any arithmetical predicate Π in $\mathscr{L}(\mathbf{PA})$—of which there are infinitely many—each of the following is an axiom:

- $\forall x \, [\Pi/x \to \forall y \, [y = x \to \Pi/y]]$.
- $[\Pi/\mathbf{0} \to [\forall x \, [\Pi/x \to \Pi/\mathbf{S}x] \to \forall x \, \Pi/x]]$.

The first of these corresponds to basic principle (1), the second to basic principle (2). And that completes the list. It also reveals **PA**'s axiomatizability.

The sufficient strength of **PA**

We can show that **PA** is sufficiently strong. This is significant not least because it assuages the concern raised in Chapter 5, that we are expecting too much of an arithmetical theory if we insist that it should meet this requirement. It is also significant because **PA**'s resources appear so meagre and we can learn a great deal from seeing how, even so, they guarantee its sufficient strength.

Very well, but how do we show that **PA** is sufficiently strong?

It would be comparatively easy to show that **PA** fits the bill as far as many specific decidable sets of natural numbers are concerned. Consider, for instance, the set of odd numbers. We could readily show that the arithmetical predicate

(*) $\forall x \neg [(x \times \mathbf{SS0}) = \_\,]$

characterizes this set and that **PA** contains all true instances of (*), such as

$\forall x \neg [(x \times \mathbf{SS0}) = \mathbf{SSS0}],$

and the negations of all false instances of (*), such as

$\forall x \neg [(x \times \mathbf{SS0}) = \mathbf{SSSS0}].$

But how do we show that something similar applies to *every* decidable set of natural numbers?

It is unfortunately beyond the scope of this book to answer this question in any detail. But there is one point that must be emphasized: we cannot hope to show any such thing unless and until the informal concept of decidability has been replaced by something more precise and more formal.

By what, though?

Several very different candidates have been proposed. And here we come upon a remarkable mathematical fact. Diverse as they are, these candidates turn out to be demonstrably equivalent. This suggests that they are different ways of directing attention to the same important mathematical phenomenon—and that any of them would be fit to do the work required here. I will provide a sketch of three of them.

(i) *Turing computability*: The first candidate exploits the idea that a set of natural numbers is decidable if and only if a computer could determine whether any given natural number belongs to the set. This idea is made precise in terms of a certain simple kind of computer known as a Turing machine. Turing machines are named after the great pioneer of computation Alan Turing who devised the concept. And a set of natural numbers E is said to be Turing computable when a Turing machine could determine whether any given natural number belongs to E.

(ii) *General recursiveness*: The second candidate takes for granted that certain basic numerical operations are algorithmic and that certain ways of combining numerical operations are likewise algorithmic. An operation is said to be general recursive when it can be obtained from these basic operations by these methods of combination. And a set of natural numbers E is said to be general recursive when there is a general recursive operation such that, if its input is a natural number belonging to E, then its output is 0, and if its input is a natural number not belonging to E, then its output is 1. (It is as though specifying a natural number n as input represents asking whether or not n belongs to E, and 0 represents the answer *yes*, 1 the answer *no*.)

(iii) *Recursiveness by finite axiomatization*: The third candidate fastens on the idea that the decidability of a set of natural numbers involves finitude—either the finitude of an account of what it takes for any natural number to belong to the set or, equivalently, the finitude of a description of an algorithm for determining whether any natural number belongs to the set. A set of natural numbers E is said to be recursive by finite axiomatization when there is a finite and consistent set of statements belonging to some formal language that can serve as axioms for proving either that a given natural number belongs to E or that it doesn't as the case may be.

The thesis that any of these candidates can be regarded as a suitably precise formal counterpart of the informal concept of decidability, as applied to sets of natural numbers, is known as Church's thesis, having been famously mooted by the logician Alonzo Church. It is also sometimes known as the Church–Turing thesis, since it was also mooted by Turing. It is nowadays routinely taken for granted. In what follows I will presuppose it, along with the corresponding adaptation of the definition of sufficient strength.

Church's thesis gives us license to work with a precise formal concept in place of the informal concept of decidability as applied to sets of natural numbers. It also, indirectly, gives us license to work with a precise formal concept in place of the informal concept of decidability as applied to sets of expressions in a formal language. For we can exploit a Gödel numbering. We can count a set of expressions as having the requisite precise property when the corresponding set of natural numbers has the corresponding precise property. Admittedly, this doesn't completely eliminate the informality: the Gödel numbering itself needs to be algorithmic, and our understanding of *that* remains informal. But it does eliminate a further significant element of informality.

Incidentally, although we can show that **PA** is sufficiently strong, we do not in fact *need* to do so for current purposes. We need only show, as Gödel himself did in the case of his own sample theory, that **PA** satisfies a somewhat weaker condition. All that matters, as we'll see, is that **PA** contains statements that correspond in a suitable way, *via* some Gödel numbering, to statements about its own constitution. That will be enough for us to establish its incompleteness. Any theory that is sufficiently strong contains such statements; but there are theories that contain such statements without being sufficiently strong.

The soundness of **PA**

PA is axiomatizable and sufficiently strong. Is it sound?

This takes us into philosophical territory as much as it does into mathematical territory. I touched on some of the issues at the end of Chapter 2. Rather than revisiting those issues now, let us for the time being just *assume*, what in any case seems intuitively correct, that **PA** is not only axiomatizable and sufficiently strong, but also sound. (Those axioms all look true!)

It follows that **PA** is consistent. It also follows that **PA** has a more complex and stronger property known as ω-consistency, which will be important later and which (like consistency) can be understood in non-semantic terms. How is ω-consistency defined? It is easier in fact to define its opposite, ω-inconsistency, and then simply to call a theory ω-consistent when it is not ω-inconsistent.

> **ω-inconsistency**: A theory T will be said to be ω-inconsistent when there is an arithmetical predicate Π in $\mathscr{L}(T)$ such that T contains:
> - $\Pi/\mathbf{0}$;
> - $\Pi/\mathbf{S0}$;
> - $\Pi/\mathbf{SS0}$;
> - $\Pi/\mathbf{SSS0}$;
>
>   .
>   .
>   .
>
> - $\neg\, \forall x\, \Pi/x.$

Less formally, a theory T will be said to be ω-inconsistent when there is some property P such that T includes statements to the effect that 0 possesses P, that 1 possesses P, that 2 possesses P, that 3 possesses P, and so on, but also that not every natural number possesses P; it will be said to be ω-consistent otherwise.

A sketch of Gödel's proof

I will now indicate how to construct a statement s in $\mathscr{L}(\mathbf{PA})$ such that neither s nor $\neg\, s$ belongs to $\mathbf{PA}$, in other words a statement s in $\mathscr{L}(\mathbf{PA})$ that testifies to $\mathbf{PA}$'s incompleteness. My construction will differ somewhat from Gödel's, but only in inconsequential ways designed to highlight the connections between Gödel's proof and the diagonal proof in Chapter 5.

Let us first consider an infinite table τ^+ of *yeses* and *noes* akin to the tables τ and τ^* considered in Chapter 5. As before, this will involve a Gödel numbering for $\mathscr{L}(\mathbf{PA})$—call it $\mathbf{G^+}$—that will allow us to list all the arithmetical predicates in $\mathscr{L}(\mathbf{PA})$:

Π^+_0
Π^+_1
Π^+_2
.
.
.

The *yeses* and *noes* in τ^+ register, as in the non-semantic version of the diagonal proof, not whether or not successive arithmetical predicates are true of successive natural numbers, but whether or not successive instances of successive arithmetical predicates belong to $\mathbf{PA}$. More precisely, given any arithmetical predicate Π^+_n, and given any natural number m, there is a *yes* at the point in τ^+ where the row for Π^+_n meets the column for m if the statement $\Pi^+_n(m)$ belongs to $\mathbf{PA}$, and there is a *no* there otherwise. (Less precisely, the *yeses* and *noes* in τ^+ register, not whether or not successive arithmetical predicates are true of successive natural numbers, but whether or not **PA** 'says' they are.)

As before, we can identify a set D^+ containing all and only those natural numbers n with a *no* at the relevant point on the diagonal of τ^+, that is to say the point in τ^+ where the row for Π^+_n meets the column for n. Suppose, by way of illustration, that $\Pi^+_{5,098}$ turns out to be the arithmetical predicate (*) considered earlier which characterizes the set of odd numbers. Then the statement $\Pi^+_{5,098}(5,098)$ is false, since 5,098 is not odd. So, given that all the statements belonging to **PA** are true, there is a *no* at the point on the diagonal where the row for $\Pi^+_{5,098}$ meets the column for 5,098. So 5,098 belongs to D^+.

Now, can we do as we did at the analogous stage in the diagonal proof and argue that there is an arithmetical predicate Π^+_i in $\mathscr{L}(\mathbf{PA})$ that characterizes D^+?

We can. *However*—and this is crucial—we can't do so by mimicking what we did before and arguing for this by direct appeal to **PA**'s sufficient strength. For there is no reason to think that D^+ shares with either of its counterparts D and D^* considered in Chapter 5 the feature of being decidable. This is because there is no reason to think that **PA** itself is decidable. (There is no analogue of the arguments in Chapter 5 for the decidability of the theories **T** and **T***, which traded on their completeness.)

Now **PA**'s sufficient strength does in fact ultimately guarantee the existence of such an arithmetical predicate Π^+_i. Indeed, as I pointed out earlier, it more than guarantees it. But this is something that has to be shown—as, of course, does the fact that **PA** is sufficiently strong in the first place. (**T** and **T*** were simply *given* as such.) So it's impossible to evade significant spadework at this stage in the proof. We have no alternative but actually to *construct* the arithmetical predicate Π^+_i. A full account of its construction lies beyond the scope of this book (although I will go into a little more detail in the next section). Some of the technical details are very intricate. Even so, there is a sense in which this part of the proof is, relative to the proof as a whole, sheer slog.

Once $\mathbf{\Pi}^{+}_{i}$ has been constructed, we have an arithmetical predicate which, if it were written out in primitive notation, would be unimaginably complex. No human being would be able to survey it, still less to make any sense of it. That does not matter. (Many significant mathematical expressions are like that. That is why mathematicians rely so heavily on the use of defined terms as abbreviations.) What matters is the rôle that $\mathbf{\Pi}^{+}_{i}$ plays in relation to $\mathbf{\tau}^{+}$. And we can ask, much as we did in the diagonal proof, what happens at the point on the diagonal of $\mathbf{\tau}^{+}$ where the row for $\mathbf{\Pi}^{+}_{i}$ meets the column for i.

It was the corresponding question that gave rise to contradiction in the diagonal proof. There was no satisfactory answer. (This is really just another way of saying that the corresponding arithmetical predicate could not be any of those on the corresponding list. Its row differed from each of theirs at the point at which the latter met the diagonal: thus it differed from the first row on the list in its first place, from the second in its second place, and so on.) In the *current* proof, however, there is no analogous contradiction. Reversing all the *yeses* and *noes* on the diagonal certainly produces a sequence that is not the same as any row on the list. But there is no longer any reason to think that it *should* be the same as any of them. In particular, there is no longer any reason to think that it should be the same as the row for $\mathbf{\Pi}^{+}_{i}$, even though $\mathbf{\Pi}^{+}_{i}$ characterizes $\mathbf{D}^{+}$. (Recall that the *yeses* and *noes* in $\mathbf{\tau}^{+}$ register whether **PA** 'says' that something is so, not whether it is so.) Nor therefore is there any reason to think that there is a problem about what happens at the point on the diagonal where the row for $\mathbf{\Pi}^{+}_{i}$ meets the column for i.

Admittedly, there cannot be a *yes* there. For suppose there were. This would mean that $\mathbf{\Pi}^{+}_{i}(i)$ belonged to **PA**. However, it would also mean, given the very definition of $\mathbf{D}^+$, that i did not belong to $\mathbf{D}^+$. So, given that $\mathbf{\Pi}^+_i$ characterizes $\mathbf{D}^+$, $\mathbf{\Pi}^+_i(i)$ would be false. And for a false statement to belong to $\mathbf{PA}$ is precluded by $\mathbf{PA}$'s soundness.

But there is no reason why there shouldn't be a *no* at the given point on the diagonal. If there were, this would mean that $\mathbf{\Pi}^+_i(i)$ did not belong to $\mathbf{PA}$. It would also mean, given the very definition of $\mathbf{D}^+$, that i did belong to $\mathbf{D}^+$. So, given that $\mathbf{\Pi}^+_i$ characterizes $\mathbf{D}^+$, $\mathbf{\Pi}^+_i(i)$ would be true. *But this is not precluded by anything.* There is no threat to $\mathbf{PA}$'s soundness in saying that it does not contain a given true statement in its language. $\mathbf{PA}$ may not contain the negation of that statement either. $\mathbf{PA}$, unlike either $\mathbf{T}$ or $\mathbf{T}^*$, may be incomplete.

And that is the point. Granted $\mathbf{PA}$'s soundness, we can conclude that $\mathbf{\Pi}^+_i(i)$ is true and does not belong to $\mathbf{PA}$. But we can also conclude that its negation $\neg\,\mathbf{\Pi}^+_i(i)$, which is false, does not belong to $\mathbf{PA}$ either, on pain of violating $\mathbf{PA}$'s soundness. So, since neither $\mathbf{\Pi}^+_i(i)$ nor $\neg\,\mathbf{\Pi}^+_i(i)$ belongs to $\mathbf{PA}$, $\mathbf{\Pi}^+_i(i)$ is the statement we were aiming to construct. *Q.E.D.*

A closer look at the construction of $\mathbf{\Pi}^+_i(i)$

Let us look a little more closely at the construction of $\mathbf{\Pi}^+_i(i)$, with a view to understanding better what it comes to. We already have some idea of what it comes to in the fact that, in the sketch of Gödel's proof just given, we were able to see that either $\mathbf{\Pi}^+_i(i)$ belongs to $\mathbf{PA}$ and is false or it does not belong to $\mathbf{PA}$ and is true; in other words, that $\mathbf{\Pi}^+_i(i)$ is true if and only if it does not belong to $\mathbf{PA}$. So it is *as if* $\mathbf{\Pi}^+_i(i)$ asserts, of itself, that it does not belong to $\mathbf{PA}$. One of my aims in this section is to elucidate this 'as if'.

Very well, how, in outline, is $\mathbf{\Pi}^+_i(i)$ constructed?

Let me begin with some points of terminology. I will follow the practice of Chapter 5 and call the natural number assigned by the

Gödel numbering $\mathbf{G^+}$ to any expression or proof its $\mathbf{G^+}$-number. I will call a proof using **PA**'s axiomatization a **PA**-proof. And I will say that a statement in $\mathscr{L}(\mathbf{PA})$ that has a **PA**-proof is **PA**-provable, while a statement in $\mathscr{L}(\mathbf{PA})$ that lacks a **PA**-proof is **PA**-unprovable. Note that a statement is **PA**-provable if and only if it belongs to **PA**.

Now consider the following definition:

> **Numerical-diagonalization**: Numerical-diagonalization is the operation which, given any natural number n as input, yields the $\mathbf{G^+}$-number of $\mathbf{\Pi^+}_n(n)$ as output. This output will be said to be the numerical-diagonal of n.

Numerical-diagonalization is an algorithmic operation that takes one natural number as input and yields another as output. In that respect, it differs only in its complexity from such operations as doubling, or squaring, or that boring operation that I mentioned in Chapter 1 which, given any natural number as input, yields the cube of the next largest prime number as output. In particular, numerical-diagonalization can be defined in purely arithmetical terms—the terms available in $\mathscr{L}(\mathbf{PA})$—although the definition would be eye-wateringly complex. Were *we* to calculate, say, the numerical-diagonal of 7, we would not directly appeal to this definition. We would first determine which arithmetical predicate is $\mathbf{\Pi}_7$, or equivalently which arithmetical predicate has the eighth lowest $\mathbf{G^+}$-number. We would then replace each occurrence of the ellipsis '_' in $\mathbf{\Pi}_7$ by **SSSSSSS0**, that is we would construct the statement $\mathbf{\Pi}_7(7)$. We would then calculate the $\mathbf{G^+}$-number of this statement. But the procedure we would thereby follow could in principle be mimicked by a procedure involving nothing but steps of arithmetic. That is what Gödel showed.

Now consider this definition:

> **Numerical-proof**: Numerical-proof is the relation that obtains between the $\mathbf{G^+}$-number of a **PA**-proof and the $\mathbf{G^+}$-

number of the statement proved. When a given natural number n bears this relation to a given natural number m, n will be said to numerically-prove m.

Given the algorithmic nature of proofs, numerical-proof is another relation that can be defined in purely arithmetical terms. This is something else Gödel showed. It follows that there is an arithmetical predicate in $\mathscr{L}(\mathbf{PA})$ that characterizes the set of $\mathbf{G}^{+}$-numbers of statements that are $\mathbf{PA}$-provable. This is a formal version of 'There is a natural number x such that x numerically-proves _'.

By the same token there is an arithmetical predicate in $\mathscr{L}(\mathbf{PA})$ that characterizes the set of natural numbers that are *not* $\mathbf{G}^{+}$-numbers of statements that are $\mathbf{PA}$-provable: a formal version of 'There is *no* natural number x such that x numerically-proves _', or '$\forall x\, x$ does not numerically-prove _'.

We can now put the pieces together. We can construct an arithmetical predicate in $\mathscr{L}(\mathbf{PA})$ that is a formal version of '$\forall x\, x$ does not numerically-prove the numerical-diagonal of _' And this, as we can see if we think about it, is $\mathbf{\Pi}^{+}_{i}$. (Or rather, it *can* be $\mathbf{\Pi}^{+}_{i}$. There is more than one arithmetical predicate that characterizes $\mathbf{D}^{+}$. This is one of them.) The statement $\mathbf{\Pi}^{+}_{i}(i)$ is then a formal version of '$\forall x\, x$ does not numerically-prove the numerical-diagonal of i'.

> $\mathbf{\Pi}^{+}_{i}(i)$: $\forall x\ x$ does not numerically-prove the numerical-diagonal of i.

But what does this come to? Well, what is the numerical-diagonal of i? By definition, it is the $\mathbf{G}^{+}$-number of $\mathbf{\Pi}^{+}_{i}(i)$. That is, it is the $\mathbf{G}^{+}$-number of *this very statement*. The upshot is that, according to $\mathbf{\Pi}^{+}_{i}(i)$, no natural number numerically-proves $\mathbf{\Pi}^{+}_{i}(i)$'s own *$\mathbf{G}^{+}$-number*, which is as much as to say that $\mathbf{\Pi}^{+}_{i}(i)$ itself is $\mathbf{PA}$-unprovable, or, equivalently, that $\mathbf{\Pi}^{+}_{i}(i)$ itself does not belong to $\mathbf{PA}$.

So, to repeat, it is *as if* $\Pi^+_i(i)$ asserts, of itself, that it does not belong to **PA**. However, there also seems to be a clear sense in which it does not actually assert any such thing. To begin with, there is an issue about whether we should ever say, of any statement, that it 'asserts' anything, even when we are not eschewing the use of such semantic concepts. (Perhaps only language-users 'assert' things, *by means of* statements.) But if or in so far as $\Pi^+_i(i)$ does assert anything, then it asserts something about natural numbers. Admittedly, $\Pi^+_i(i)$ has, as one of its foci, one particular natural number which, for reasons that are quite extraneous to $\Pi^+_i(i)$ itself and that depend on an utterly arbitrary assignment of natural numbers to expressions, happens to have been assigned to a statement we happen to be interested in: $\Pi^+_i(i)$ itself. But that is incidental. (We wouldn't have had any interest in this statement if we had been using one of the infinitely many other possible Gödel numberings.) The fact that $\Pi^+_i(i)$ has this focus is no different, in principle, from the fact that Goldbach's conjecture has, as one of its foci, the natural number 2 in which we may likewise have some independent interest.

To be sure, there is a level of mathematical-cum-philosophical sophistication at which we can *identify* expressions in $\mathscr{L}(\mathbf{PA})$ with their $\mathbf{G}^+$-numbers, just as, in analytic geometry, we can identify points on a plane with pairs of real numbers (their co-ordinates) on the grounds that the latter can do the same mathematical work as the former. If we do that, then we can say that $\Pi^+_i(i)$ has, as one of its foci, itself; and that it is concerned with its own **PA**-unprovability. But we have to earn the right to say that. And this will include satisfying ourselves that the $\mathbf{G}^+$-numbers of expressions in $\mathscr{L}(\mathbf{PA})$ really can do the same mathematical work as the expressions themselves, without remainder and without contradiction. We *can* satisfy ourselves of this. But we can also rest content with the 'as if' and try to be as clear as possible about what it entails and what it does not entail.

The second theorem

I have argued that $\Pi^+_i(i)$ is true. But on what basis exactly? Given that $\Pi^+_i(i)$ is **PA**-unprovable, I must, at some point, have made use, however tacitly, of *something* that is not part of **PA**'s own axiomatization. What?

We can put the question as follows. How might the axiomatization of **PA** be supplemented for my argument for the truth of $\Pi^+_i(i)$ to be cast as a formal proof expressible in $\mathscr{L}(\mathbf{PA})$?

It turns out that a *great deal* of that proof would require no supplementation whatsoever. To appreciate how much constitutes this 'great deal', reconsider my argument that $\Pi^+_i(i)$ does not belong to **PA**, or equivalently that $\Pi^+_i(i)$ is **PA**-unprovable, or equivalently again that there cannot be a *yes* at the relevant point on the diagonal. This argument, which was a crucial component of my argument that $\Pi^+_i(i)$ is true, appealed to **PA**'s soundness. But it need not have done. It need only have appealed to something weaker and non-semantic: **PA**'s consistency. To see why, recall that the $\mathbf{G}^+$-number of $\Pi^+_i(i)$ is the numerical-diagonal of i. And suppose $\Pi^+_i(i)$ were **PA**-provable. Then some natural number would numerically-prove the numerical-diagonal of i. So, for reasons connected with **PA**'s sufficient strength, there would be a **PA**-proof of a statement to that effect, that is to the effect that some natural number numerically-proves the numerical-diagonal of i. We can be more specific. Consider $\neg\,\Pi^+_i(i)$. Precisely what this is is a statement to that effect; for it is the negation of $\Pi^+_i(i)$, and, as such, it is a denial that *no* natural number numerically-proves the numerical-diagonal of i. And there would be a **PA**-proof of *that*. It follows that there would be a **PA**-proof of both $\Pi^+_i(i)$ and $\neg\,\Pi^+_i(i)$. **PA** would be inconsistent.

Note, incidentally, that an appeal to **PA**'s consistency would not have been enough to establish the **PA**-unprovability of $\neg\,\Pi^+_i(i)$.

My argument for *this*, like my argument for the **PA**-unprovability of $\Pi^+_i(i)$, appealed to **PA**'s soundness. And, although this argument too need only have appealed to something weaker and non-semantic, what it needed to appeal to was **PA**'s ω-*consistency*—which is stronger than **PA**'s consistency. Still, an appeal to **PA**'s ω-consistency would have been enough. For, as I have indicated, ¬ $\Pi^+_i(i)$ is a denial that no natural number numerically-proves the numerical-diagonal of i. And the fact that $\Pi^+_i(i)$ is **PA**-unprovable means, for reasons once again connected with **PA**'s sufficient strength, that a formal version of each of the following is **PA**-provable:

0 does not numerically-prove the numerical-diagonal of i
1 does not numerically-prove the numerical-diagonal of i
2 does not numerically-prove the numerical-diagonal of i
3 does not numerically-prove the numerical-diagonal of i

.
.
.

So if **PA** is ω-consistent, then ¬ $\Pi^+_i(i)$ is **PA**-unprovable.

Gödel's proof was later refined by J.B. Rosser, who showed how to construct a somewhat more complicated statement s such that we can see, merely by appeal to **PA**'s consistency, that neither s nor ¬ s belongs to **PA**.

To return to $\Pi^+_i(i)$ itself, we have seen that we are in a position to affirm the following:

(1) If **PA** is consistent, then $\Pi^+_i(i)$ is **PA**-unprovable.

Now there is a statement in $\mathscr{L}(\textbf{PA})$ that corresponds to (1)—in the sense of correspondence with which we were concerned in the previous section. That is, there is a statement in $\mathscr{L}(\textbf{PA})$ such that we can show that it is true if and only if (1) is true. This is because there is a statement c in $\mathscr{L}(\mathbf{PA})$ that corresponds to the statement that $\mathbf{PA}$ is consistent—I will justify this claim shortly—and there is a statement u in $\mathscr{L}(\mathbf{PA})$ that corresponds to the statement that $\mathbf{\Pi}^+_i(i)$ is $\mathbf{PA}$-unprovable—I will justify this claim too. So, corresponding to (1) as a whole, there is the statement:

(2) $[c \rightarrow u]$.

What then is c? Well, for $\mathbf{PA}$ to be inconsistent would be for every statement in $\mathscr{L}(\mathbf{PA})$ to be $\mathbf{PA}$-provable. (The axiomatization of $\mathbf{PA}$, like the axiomatizations considered in Chapter 3, incorporates a formal equivalent of the principle: 'Believe that and you'll believe anything'.) So we can register that $\mathbf{PA}$ is consistent by saying that the false statement $\mathbf{0} = \mathbf{S0}$, say, is $\mathbf{PA}$-unprovable. But $\mathbf{0} = \mathbf{S0}$ has a $\mathbf{G}^+$-number. Let us suppose, for the sake of argument, that the $\mathbf{G}^+$-number of $\mathbf{0} = \mathbf{S0}$ is 9,450. (This is what it would be if the $\mathbf{G}^+$-numbers of $\mathbf{0}$, $\mathbf{S}$, and $=$ were 1, 2, and 3, respectively, and if $\mathbf{G}^+$ were based on the prime-number method of calculating assignments which I described at the end of Chapter 4. For, in that case, the $\mathbf{G}^+$-number of $\mathbf{0} = \mathbf{S0}$ would be $2^1 \times 3^3 \times 5^2 \times 7^1$, which is 9,450.) Then c is a formal version of '$\forall x$ x does not numerically-prove 9,450'.

What about u? Well, as we showed in the previous section, *u is simply $\mathbf{\Pi}^+_i(i)$ itself.* So (2) can be rewritten as follows:

(2) $[c \rightarrow \mathbf{\Pi}^+_i(i)]$.

Now—and this too is crucial, although demonstrating it is again unfortunately beyond the scope of this book—(2) is $\mathbf{PA}$-provable. (This is the 'great deal' to which I referred earlier.)

It follows, given that $\mathbf{\Pi}^+_i(i)$ is $\mathbf{PA}$-*un*provable, that c must be $\mathbf{PA}$-unprovable too. Otherwise, the $\mathbf{PA}$-proof of c could be combined with that of (2) to yield a $\mathbf{PA}$-proof of $\mathbf{\Pi}^+_i(i)$.

So there we have it. This statement c, which corresponds to a statement of **PA**'s consistency, does not itself admit of a **PA**-proof and is the extra basic principle required to prove $\Pi^+_i(i)$. (We could of course add c as a new axiom, to generate a new theory containing $\Pi^+_i(i)$, but then Gödel's entire proof could be applied to this new theory, which would suffer from an analogous incompleteness of its own.) Acknowledging **PA**'s consistency enables us to prove things that are **PA**-unprovable. But we needn't have any qualms about acknowledging **PA**'s consistency—not unless we have qualms about acknowledging its soundness, of which its consistency is a straightforward corollary.

And it is a suitably generalized version of the claim that **PA**'s axiomatization cannot be used to prove c that is known as Gödel's second theorem:

> No consistent, sufficiently strong, axiomatizable theory can contain a statement corresponding to a statement of its own consistency.

Chapter 7
Hilbert's programme, the human mind, and computers

I will use these final two chapters to explore some of the philosophical implications of Gödel's theorem.

Threats to Hilbert's programme

I begin with the implications of the theorem for Hilbert's programme. Recall the distinction that Hilbert drew between two kinds of mathematical statement: those that he called 'finitary', which he argued were concerned with the concatenation and manipulation of signs; and those that he called 'ideal', which he argued were meaningless devices to facilitate proofs. Only the former, in Hilbert's view, were true or false. Arithmetical statements, he thought, could be of either kind. Hilbert's programme involved: (i) devising an axiomatization **H** powerful enough to enable us to prove all finitary truths; and (ii) providing reassurance that **H** was not so powerful that it enabled us to prove some finitary falsehoods as well. Providing this reassurance could not consist in simply appealing to the truth of each of the axioms in **H**, since some of these axioms would be ideal statements and therefore *not* true. Rather, it would require proving the consistency of **H**. Moreover, it would require doing so without exploiting any of the ideal statements at issue, or any other ideal statements for that matter, since this would obviously just require a similar reassurance of its own. The proof of **H**'s consistency would have to be purely finitary.

9. David Hilbert.

Gödel's theorem certainly threatens the execution of (ii). It even
threatens the execution of (i). I should concede, before
proceeding, that Hilbert would not have accepted everything in

the proofs of the theorem as presented in the previous two chapters. For instance, he would not have accepted the crucial assumption underpinning the semantic version of the diagonal proof: that any statement in any given formal language is either true or false. Still, there was enough here that he could and did accept to cause problems for him.

Why is the execution of (i) threatened? Because what Hilbert seemed to be envisaging, at least as a paradigm, was an axiomatized theory that was not only consistent but sufficiently strong and complete. That is, he seemed to be envisaging a theory that included, in addition to all the finitary statements that counted as true by his lights, all the ideal statements that counted as true by more traditional lights. (His aim was not to challenge ideal mathematics. His aim was to vindicate it. What he wanted to challenge was a certain natural way of conceiving it.) But Gödel's theorem shows that nothing matches this paradigm. Any axiomatized theory that is consistent and sufficiently strong is incomplete.

And why is the execution of (ii) threatened? Because Gödel's second theorem seems to entail that, whatever form **H** took, no proof of its consistency could be recast and implemented using **H** itself; therefore that no proof of its consistency could avoid using an axiomatization that was at least as powerful as **H**; and therefore that no proof of its consistency could be purely finitary.

Take **PA**. Although the theory axiomatized by **H** would have to include more than that, it would have to include at least that (or something like that). So it is instructive to look specifically at how these considerations bear on **PA**. I said in Chapter 6 that we needn't have any qualms about acknowledging the consistency of **PA**—unless we have qualms about acknowledging its soundness. Hilbert *would* have had such qualms. He would not have been prepared to grant that every statement in **PA** is true. To satisfy ourselves that **PA** is consistent, Hilbert would have said, we must appeal only to finitary truths about statements in $\mathscr{L}(\mathbf{PA})$—truths of a kind that, via some Gödel numbering, correspond to the finitary truths about natural numbers that **PA** itself contains—and then prove that one such statement, say the finitary falsehood **0 = S0**, does not belong to **PA**. But this is precisely what Gödel's second theorem seems to entail we cannot do.

Admittedly, none of this constitutes a decisive refutation of Hilbert's programme. This is because of lingering unclarity in how the terms 'finitary' and 'ideal' are being understood. Gödel himself said as much in his article:

> I wish to note expressly that [my second theorem does] not contradict Hilbert's . . . viewpoint. For this viewpoint presupposes only the existence of a consistency proof in which nothing but finitary means of proof is used, and it is conceivable that there exist finitary proofs that *cannot* be expressed in the [relevant] formalism.

That is, there may be rules that can legitimately be classified as finitary, that are nevertheless not part of **H**, and that we can use to provide a finitary consistency proof of **H**. If, furthermore, the ideal part of **H** needn't answer to any demand but the demand for effectiveness—if, in particular, it needn't answer to the demand for *completeness*—then it is still possible to cherish the hope of executing both (i) and (ii).

Even so, this is rather cold comfort for any defender of Hilbert's programme, not least because it exploits what ought really to be seen as a *problem* with the programme: an obscurity in what exactly Hilbert was envisaging. Moreover, a delicate balance now needs to be struck. A precise explanation of 'finitary' is required that *both* averts the threat posed by Gödel's theorem *and* satisfies the original rationale of the label: that it entitles us to say that only when a mathematical statement is finitary does it have whatever it takes to be classified as true or false.

Self-consciousness

Part of the problem for Hilbert lies in the fact that, when we accept a mathematical axiomatization A, this provides us with two quite distinct ways of achieving mathematical insights. We can use A to prove things. Or we can reflect self-consciously on the nature of A itself. In particular, in certain favourable cases, we can acknowledge that A is sound; conclude that A is consistent; formalize this as a new basic principle about the nature of mathematical reality; and proceed to prove new things. But we have to be clear about what would have troubled Hilbert here. What would have troubled him is not the mere thought that we can achieve mathematical insights by reflecting self-consciously on some mathematical axiomatization that we accept, nor even that we can do this by acknowledging that the axiomatization is sound. What would have troubled him is the thought that we can do these things when part of the axiomatization is ideal. For that would mean acknowledging that the axiomatization as a whole faithfully depicts some infinite tract of mathematical reality. It would mean acknowledging the truth of certain ideal statements. *That* would have troubled Hilbert.

Still, we surely *can* achieve mathematical insights by reflecting self-consciously on mathematical axiomatizations that we accept. In particular, we can do this in the case of arithmetical axiomatizations. We can stop thinking directly about natural numbers and start thinking instead about (our own) *thinking* about natural numbers and about how we have axiomatized it. Granted Gödel's theorem, this can give us new insights into what the natural numbers themselves are like.

The Lucas–Penrose argument

There is an argument that has been advanced on the strength of these considerations for the conclusion that Gödel's theorem shows the human mind to have mathematical powers beyond those of any possible computer. The two best known proponents of this argument are John Lucas and Roger Penrose, and I will refer to the argument as the Lucas–Penrose argument.

For the sake of simplicity, I will revert to the anti-Hilbertian view that every statement in $\mathscr{L}(\textbf{PA})$ is either true or false, though this is not strictly necessary for the argument. With that simplification, the argument can be presented as follows.

Imagine a computer C that can tell that certain statements in $\mathscr{L}(\textbf{PA})$ are true. How so? There must be some sound, finite stock of basic principles and rules for arithmetic that have somehow been incorporated into C's software. That is, C must somehow have been programmed to exploit some sound axiomatization A of arithmetic. Quite *how* it exploits A is a further matter. Perhaps, if 'asked' about one of the truths in question, it searches systematically through proofs that use A until it discovers a proof of that truth. Or perhaps it is never 'asked' about anything: perhaps it is searching through these proofs anyway, simply churning out all the statements proved. But, however C exploits A, the set T of statements in $\mathscr{L}(\textbf{PA})$ that C can tell are true is just the theory axiomatized by A.

It follows from Gödel's theorem that there are truths in $\mathscr{L}(\textbf{PA})$ that C cannot tell are true. Indeed this follows even if we grant C enough mathematical power for T to be sufficiently strong.

Now comes the crucial point. *We*—human beings—can always outwit C. Given any such limited and axiomatizable set of truths T, *we* can always recognize the soundness, and therefore the consistency, of T. And, for reasons implicit in the proof of Gödel's theorem, this means that we can always recognize the truth of certain statements in $\mathscr{L}(\textbf{PA})$ that do not belong to T. (Of course, if T were *not* sufficiently strong, then we might not need to be that sophisticated. For instance, if T contained only statements about

99

addition, then we could just turn our attention to statements about multiplication. But the point is, being that sophisticated would always be an option.)

Given any possible computer, then, the set of truths in $\mathscr{L}(\mathbf{PA})$ that it can recognize as such is always surpassed by the set of truths in $\mathscr{L}(\mathbf{PA})$ that we—human beings—can recognize as such. Moreover, the considerations in the previous section seem to indicate why: what we can do, which no computer can do, is to indulge in self-conscious reflection on any algorithmic representation of any of our mathematical powers in a way that involves the exercise of even greater mathematical powers. Indeed, there is no reason as yet to suppose that *we* cannot tell that *every* true statement in $\mathscr{L}(\mathbf{PA})$ is true—although, if we can, it may remain a mystery how we can, just as it may remain a non-trivial matter to exercise the ability.

Such is the Lucas–Penrose argument. Those who subscribe to it often parade it as one of the main reasons why we do not have to see Gödel's theorem in a negative light, despite the fact that it draws attention to certain inherent limitations to what we can achieve in mathematics. Instead, they say, we should celebrate the inexhaustible powers of the human mind to which it draws attention. There is nothing that we can recognize as a good axiomatization of arithmetic such that we cannot recognize something else as an even better one.

My own view, as I will indicate in Chapter 8, is that there are many other reasons why we do not have to see Gödel's theorem in a negative light. But in any case, as we'll see in the next section, the Lucas–Penrose argument is vulnerable to attack.

Responses to the Lucas–Penrose argument

The Lucas–Penrose argument has provoked much discussion. More specifically, it has provoked much opposition.

There is one objection to the argument that I want to set aside straight away. I have in mind the objection that, provided the identity of any given computer C does not depend even partially on its software, and provided we are allowed to discount any limitations arising from its hardware, then C can always do more than its current software enables it to do—simply through a suitable update of that software. That objection makes the view to which the Lucas–Penrose argument is opposed trivial, and it deprives the argument itself of any interest. For, patently, given any true statement s in $\mathcal{L}(\mathbf{PA})$, C can tell that s is true if its software is suitably updated: at the limit C can be programmed to use an axiomatization that includes s as an axiom. So the only interesting question—and it *is* an interesting question, because it makes us think not just about *which* truths can be recognized as such but about *how* they can be recognized as such—is what mathematical powers C has if its software is fixed. I will therefore construe the Lucas–Penrose argument as targeted at that question.

Still there are objections to the argument. Twin pressures have been brought to bear on it, as it were from below and from above.

The pressure from below derives from the concern that the argument downgrades computers. Is it true that the only way in which C can tell that certain statements in $\mathcal{L}(\mathbf{PA})$ are true is by exploiting a single axiomatization of arithmetic? Couldn't C be programmed to periodically replace any axiomatization A that it has been using with a more powerful one—perhaps by adding a statement corresponding to the statement that A is consistent, or perhaps in some much more recondite way?

Admittedly, this would itself require some suitable algorithm, so there is an issue about whether Gödel's theorem could be adapted to show that C was still unable to recognize the truth of certain statements in $\mathcal{L}(\mathbf{PA})$ whose truth *we* were able to recognize by reflecting on the algorithm. It is far from obvious, however, that the theorem could be adapted in this way. For the algorithm would not be *straightforwardly* arithmetical. It would provide a way of generating axiomatizations of arithmetic; it would not involve the implementation of one.

The fact that the algorithm would not be straightforwardly arithmetical suggests another possibility that we need to consider. Couldn't C follow some algorithm that achieved what was required, but *at a sub-arithmetical level*? Suppose, for instance, that C listed various arithmetical truths, including '$7 + 5 = 12$'. Couldn't it have achieved this, not by directly proving, on the strength of some basic principles and rules for arithmetic, that $7 + 5 = 12$, but rather by carrying out a series of horrendously complicated sub-processes that simply had the effect of producing that particular array of symbols on its screen—if only as a result of (ingenious) programming on the part of human beings?

So much for the pressure from below. The pressure from above derives from the concern that the argument upgrades human beings. Suppose it is true that, given any sound axiomatization A of arithmetic that C uses, we—human beings—can specify a statement in $\mathscr{L}(\mathbf{PA})$ whose truth we can recognize but that cannot be proved using A. Even so, can we be sure that, in any relevant sense, we can be 'given' all such axiomatizations? Couldn't C exploit an axiomatization that was so complex that we—human beings—could not be 'given' it? And couldn't this put C in just as good a position to tell that certain statements in $\mathscr{L}(\mathbf{PA})$ were true as we are? Perhaps there are limitations to our own mathematical powers for which the Lucas–Penrose argument makes no allowance. Perhaps the set of truths in $\mathscr{L}(\mathbf{PA})$ that we can recognize as such is in fact axiomatizable, but the axiomatization is so complex that we never make conscious use of it and indeed never could. Gödel himself had a related thought: 'There may exist…a theorem-proving machine which in fact *is* equivalent to mathematical intuition, but cannot be *proved* to be so'.

At this point advocates of the Lucas–Penrose argument are liable to object that there is a crucial difference, to which I have already referred more than once in this book, between what is possible in principle and what is possible in practice, and that it is only the former that is relevant here. For of course we cannot be 'given' all such axiomatizations *in practice*. *In practice* we can only ever be 'given' something suitably manageable. But, the objection runs, we *can* be given all such axiomatizations in principle; and, given one of them, we *can* in principle see (past) its limitations.

But there is a response to this objection. While the distinction between what is possible in principle and what is possible in practice is certainly important, and while it is certainly relevant to the significance of Gödel's theorem, it cannot be allowed application to our mathematical powers in this way unless it is also allowed application to C's. Who knows but that, *in principle*, C can tell the truth or falsity of any statement in $\mathscr{L}(\mathbf{PA})$—if only by brute force? Consider, for instance, Goldbach's conjecture. Why, in principle, shouldn't C determine whether or not this is true by considering successive even numbers greater than 2 and ascertaining in each case whether or not it is the sum of two primes? It is of no avail to protest that this procedure might never end. Something needs to be said to forestall the reply that, if C is constructed in such a way that it spends half an hour considering the first even number greater than 2, a quarter of an hour considering the second, an eighth of an hour considering the third, a sixteenth of an hour considering the fourth, and so on, then the procedure will end in (at most) an hour. To be sure, there are all sorts of reasons, including reasons deriving from the most fundamental laws of physics, why this is a mere pipe dream and not possible in any *practical* sense. But that is precisely beside the point.

My own view is that there are no really interesting conclusions to be drawn from Gödel's theorem about the relationship between the human mind and computers. Like some of what we witnessed towards the end of Chapter 1, the Lucas–Penrose argument is an attempt to locate the significance of Gödel's theorem other than where it really lies. In the next chapter, I will try to do better.

Chapter 8
Making sense in and of mathematics

The primary significance of Gödel's theorem lies simply in the intrinsic interest of the result itself together with the ingenuity and craftsmanship that are displayed in its proof. We do not need to seek repercussions elsewhere to understand why the theorem commands attention. That said, it does have repercussions elsewhere. In particular, it raises some profound questions about how we make sense in and of mathematics. It is to these questions that I turn in this chapter.

As in my presentation of the Lucas–Penrose argument in Chapter 7, I will assume that every statement in $\mathscr{L}(\mathbf{PA})$ is true or false. This is helpful for framing the issues; it is not essential to them.

Acknowledging consistency

In Chapter 6, I showed how, if we acknowledge the consistency of **PA**, we can argue for the truth of statements in $\mathscr{L}(\mathbf{PA})$ beyond those that belong to **PA** itself. In Chapter 7 we considered what it takes to acknowledge the consistency of **PA**. I suggested that we can do this by self-consciously reflecting on **PA**'s soundness. This involves not merely accepting **PA**, but accepting our *entitlement* to accept **PA**.

To amplify, there are three things at stake here:

(A_1) our acceptance of **PA**;
(E_1) our entitlement to accept **PA**, that is our entitlement to (A_1);
(A_2) our acceptance of **PA**'s consistency.

(A_1) is manifest in our actual use of **PA** to prove various statements about natural numbers. (E_1) is a feature of (A_1) that is grounded in the fact that **PA** is sound. (A_2) is something that we can achieve by self-conscious reflection on (E_1).

But there is a fourth thing at stake which we also need to consider:

(E_2) our entitlement to accept **PA**'s consistency, that is our entitlement to (A_2).

It is tempting to think that (E_2) is a feature of (A_2) that is grounded in the same thing as (E_1), that is in the fact that **PA** is sound, and consequently that (E_1) itself guarantees (E_2). But it doesn't. (E_2) goes beyond (E_1), in two significant respects. First, there are certain concepts that we need to possess in order to have (E_2) but not in order to have (E_1); concepts that we need to possess in order to achieve (A_2) in the first place. The most obvious example is the concept of consistency itself. Another, if we are to infer **PA**'s consistency from its soundness, is that of truth. Second, we cannot have (E_2) unless we have considered the axiomatization of **PA** as a whole. This is not true of (E_1). We can use **PA** to prove various statements about natural numbers, and be entitled to do so, without at any point having to consider anything other than whatever (limited) part of **PA**'s axiomatization is relevant to the proof of whatever statement we are proving. If we claim that **PA** is consistent, on the other hand, then we need to be able to rule out *any* part of its axiomatization's being incompatible with any other. Relatedly, there may be doubts about individual parts of **PA**'s axiomatization that are mild enough, in isolation, not to threaten

(E_1), but that are severe enough, in combination, to threaten (E_2). So work is still required to assure us that we really *can* acknowledge **PA**'s consistency, or, what comes to the same thing, that we can be said to know that **PA** is consistent.

Can we be said to know this? We certainly think it. But we must obviously exercise caution here. Frege, for all his logical and mathematical brilliance, mistakenly thought that his axiomatization **BLA** was consistent. Admittedly, we seem to have advanced considerably in our understanding of these matters since then; and indeed there are those who argue on empirical grounds that there cannot be a similar flaw in **PA**, since, if there were, someone would have discovered it by now. Some of the same people conclude, on the same empirical grounds, that we can indeed be said to know that **PA** is consistent. The grounds are shaky however. What if **PA** is flawed in a way that is extraordinarily subtle?

My own view, despite these concerns, is that we *can* be said to know that **PA** is consistent. However, I am not especially concerned to debate this issue now because, interesting though it is, it's not the key issue. The key issue is not whether the relation in which we stand to **PA**'s consistency is one of knowing, but how we get there. Never mind the connection between (E_1) and (E_2)— what about the connection between (A_1) and (A_2)? What is it about our acceptance of **PA** that brings in its train our acceptance of **PA**'s consistency?

A challenge presented by Gödel's theorem

As we'll see, the fact that our acceptance of **PA** brings in its train our acceptance of **PA**'s consistency—whatever the correct account of where talk of knowledge or entitlement is appropriate— connects with one of the challenges presented by Gödel's theorem. There may not appear to be anything the least challenging about it. Given that we not only accept **PA** but do so quite wittingly, surely it is only to be expected that we shall accept **PA**'s consistency? Perhaps it is. But therein lies the challenge.

I introduced this challenge in Chapter 2. Here is a slightly expanded version of it. We must, it seems, have been exposed to *something* that has conferred our understanding of the vocabulary of $\mathscr{L}(\mathbf{PA})$ on us. And whatever we have been exposed to must in turn capture all the truths expressible using that vocabulary, since their truth is fixed by its meaning. To put it slightly differently: if whatever we have been exposed to *didn't* capture all these truths, that is if there were some true statement s that it failed to capture, then it (whatever we have been exposed to) would be compatible with at least two interpretations of the vocabulary, one on which s was true and one on which s was false, and it would not have conferred understanding of the vocabulary on us. But surely what we have been exposed to must be something like a finite set of instructions for how to use the vocabulary, which must in turn be something like a finite stock of basic principles and rules involving the vocabulary; and precisely what Gödel's theorem establishes is that nothing of that kind *can* capture all the relevant truths. In particular, there are truths in $\mathscr{L}(\mathbf{PA})$ that the axiomatization of **PA** fails to capture, including, most notably, the statement c that corresponds to the statement that **PA** is consistent. What then has put us in a position to see that c is true? What has even ensured that we *agree* that c is true?

To be sure, there may be an answer to this question that appeals to something—call it X—that is more comprehensive than the axiomatization of **PA**. But unless X is of the same fundamental kind as the axiomatization of **PA**, that's to say unless X is something like a finite stock of basic principles and rules involving the vocabulary of $\mathscr{L}(\mathbf{PA})$, an account is still required of how it can confer understanding of the vocabulary on us, while if X *is* of the same fundamental kind as the axiomatization of **PA**, the problem repeats. And in any case, if this more comprehensive thing X were *merely* more comprehensive—if it merely involved some additional axioms, say—then simply appealing to it would not do justice to the point made at the beginning of this section. Somehow our acceptance of **PA** already *brings in its train* our acceptance of **PA**'s consistency—which in turn brings in its train our acceptance of c. Our acceptance of c feels as though it is of a piece with our acceptance of **PA**: the alternative of accepting $\neg c$ or even of remaining agnostic about c would not, we feel, make sense. Yet we do not seem able to vindicate this feeling in terms of any finite stock of basic principles and rules governing our use of the vocabulary of $\mathscr{L}(\mathbf{PA})$. What is going on here? How has whatever we have been exposed to conferred our understanding of the vocabulary of $\mathscr{L}(\mathbf{PA})$ on us? To answer that is the challenge.

Before I say more about how to address this challenge, I want to dismiss one popular view of mathematics that looks as though it might help. This is the view that what is captured by any given mathematical axiomatization *does* determine the meaning of the vocabulary involved, though this meaning may be indeterminate, leaving certain statements neither true nor false. (So this is a view that abandons the assumption I made at the beginning: that every statement in $\mathscr{L}(\mathbf{PA})$ is either true or false.) Another way to put this is that meaning and truth, in a mathematical context, are *relative to an axiomatization*. In particular, whether a given mathematical statement is true or false is a question that always needs to be relativized to some axiomatization A, and the answer is: 'The statement is true relative to A if it can be proved using A; it is false relative to A if it can be disproved using A; and it is neither if neither'. On this view, if we supplement any given axiomatization and create a new axiomatization that captures new truths (for instance, if we add c as an axiom to the axiomatization of **PA**), then, even if we are using exactly the same vocabulary, we are changing its meaning—just as we would be if, in passing from a discussion of chess to a discussion of bridge, we continued to make reference to kings and queens. It is as if we are playing a whole new game.

But that is needlessly counterintuitive. The idea that meaning and truth in mathematics should be relativized to some suitable practice of using mathematical vocabulary is one thing: the idea that they should be relativized to something as narrow as an axiomatization is quite another and quite unwarranted. The transition from accepting **PA** to accepting the statement c is a transition *within* a practice of using mathematical vocabulary. The latter *makes sense* in terms of the former. That is the point. *That* is what has to be accounted for.

Gödel's theorem leaves us with a basic challenge to meet about meaning and understanding.

Meaning and understanding (in mathematics)

Let us first reconsider whether capturing the truths in $\mathscr{L}(\mathbf{PA})$—that is, all and only the truths in $\mathscr{L}(\mathbf{PA})$—has to be understood in such a way that Gödel's theorem precludes our doing it. One way to meet the challenge would be to show that we had some way of capturing these truths other than by providing them with an axiomatization; and therefore that there could be something like a finite set of instructions for how to use the vocabulary of $\mathscr{L}(\mathbf{PA})$ that didn't take the form of a finite stock of basic principles and rules involving that vocabulary. Couldn't we, say, start with our axiomatization of **PA** and add some general precept to the effect that any set of statements in $\mathscr{L}(\mathbf{PA})$ that we accept, or can come to accept by suitable application of this very precept, is consistent? This would be a single overarching expression of our self-consciousness. It would not itself be a basic principle involving the vocabulary of $\mathscr{L}(\mathbf{PA})$. It would be something more like a recipe for generating such principles, and capable of leading us from acceptance of **PA** to acceptance of c and beyond. And mightn't it provide us with a way of capturing all and only the truths in $\mathscr{L}(\mathbf{PA})$?

If this suggestion were correct, the challenge presented above by Gödel's theorem would, it seems, be averted. But it is a real question how serious that challenge is even if the suggestion is not correct—even if we simply had *no* way of capturing all and only the truths in $\mathscr{L}(\mathbf{PA})$.

Let us reconsider the challenge. I have presented it as a challenge grounded in considerations about the meaning of mathematical vocabulary. But really it is grounded in much more general considerations about the meaning of *any* vocabulary. In particular, it is grounded in the elemental thought that the meaning of an expression is a matter of how it is used. This in turn prompts the following thought: if the meaning of an expression is to be grasped, then something about how the expression is used, accessible to someone who does not yet understand the expression, must serve to determine its meaning; and, whatever this is, it must, at least in principle, be susceptible to some finite non-question-begging characterization. (The requirement of finitude is to guarantee accessibility. The requirement of non-question-beggingness is to guard against a characterization such as 'We use the symbol "+" to stand for addition', which is true, but which leaves us none the wiser concerning how—independently of our understanding of this very symbol—we know what addition is.) To produce this finite non-question-begging characterization would be, as I will say, to *pin down* the expression's meaning. In the case of a mathematical expression, such as the expressions in $\mathscr{L}(\mathbf{PA})$, it would involve capturing all and only the truths in the formal language to which the expression belongs.

But these thoughts are misguided. It is true that the meaning of an expression is a matter of how it is used. And it is true that grasping the meaning of an expression is a matter of having access to something about how it is used. It simply does not follow, however, that we should always be able to pin down an expression's meaning.

In fact there is reason to doubt that we can ever do this. Gödel's theorem, in its own way, helps to make graphic something about meaning that is there to be acknowledged anyway. The point is this. The meaning of an expression has infinite possibilities woven into it. Any expression can be used in indefinitely many ways, for indefinitely many purposes, and to indefinitely many effects, whether literally or metaphorically, precisely or loosely, prosaically or poetically, straightforwardly or ironically, strictly or metonymically. No finite, non-question-begging characterization of anything is ever able to determine that full infinite potential. There is no legislating in advance for the possibilities of (creative) language-use that this potential offers. For example, there is no legislating in advance for the success of metaphorical uses of the expression, which may be contrived to describe situations completely unlike anything anybody has ever encountered before. Such is the open-endedness and versatility of meaning.

How then do people manage to grasp such meaning?

Well, how *do* they? They observe expressions being used. They try to see the point of the use. They try to use the expressions in the same way, under the guidance of promptings, corrections, and encouragement from others. We might think that, if there is nothing in how an expression is used to which they can have access *before* understanding it and which actually serves to determine the expression's (full infinite) meaning, none of this can help; that they will be confronted by something that strikes them as being, at best, inconclusive and, at worst, incomprehensible. And initially, perhaps, they will. But they eventually come to understand. It is true that this can seem quite mysterious. What we must do, however, is to see it as perfectly natural. People have shared interests, a shared sense of what is significant, and a shared appreciation of how things relate to one another (where these are partly innate and partly taught). As a result, people are able to respond to suitable training. They are able to see what other people are up to. They are able to understand one another. They are able to grasp what expressions mean. In a mathematical context, there is no reason why being shown some of the truths of a theory—seeing how these truths are proved and the kinds of justifications that are given for affirming them—should not give someone an indication of how to carry on, even though not all the truths of the theory have been, or could be, captured by what the person has been shown.

I hope this discussion takes some of the mystique out of Gödel's theorem, which now seems to give a technical twist to something non-technical and quite pervasive. We never grasp meaning by seeing its full infinite potential being realized. What we see is always a finite portion of that. But the meaning is still there to be discerned.

Gödel's theorem ends up bearing witness to the infinite potential of meaning, then. But it is worth noting, as a parting shot, that this is not just because of what it teaches us concerning the truths of arithmetic. It is also because of what it teaches us through the very methods employed in its proof and the ingenious use of linguistic resources that these themselves involve. Consider for instance the inventiveness of Gödel numbering, which enables claims about statements and proofs to be recast as claims about natural numbers. It is common among mathematicians to adopt the practice that I mentioned in Chapter 6, of using this as a licence for talking about statements and proofs as though they *are* natural numbers—claiming, for instance, that some particular natural number is a proof of some other. But doesn't this mean putting the words 'statement' and 'proof' to something *like* a metaphorical use, applying them to what only resemble statements and proofs, as a way of drawing attention to crucial similarities between two mathematical domains?

Just as Gödel's theorem can teach us something profound about meaning through its content, then, so too can it teach us something profound about meaning through the vehicle of that content. Not that this should come as any surprise. This is just what we might expect of this endlessly fascinating exercise in self-reflexivity. It is just one more manifestation of the fact that Gödel's theorem is one of the greatest monuments to mathematical excellence and also to mathematical creativity.

Appendix
A sketch of the proof of Gödel's theorem(s)

See the note at the end of the Preface for guidance concerning this Appendix and when to consult it.

Gödel's theorem: No axiomatization can capture all and only arithmetical truths.

Proof

Suppose there is an axiomatization that captures all and only arithmetical truths. Then there is an algorithm for deciding whether or not an arbitrary arithmetical statement is true. The algorithm is to search systematically through all the proofs that use the axiomatization until either a proof that the statement is true or a proof that its negation is true appears: one or other must eventually do so.

Now consider all the expressions in the language of arithmetic that characterize sets of natural numbers, such as '_ is odd' or '_ is prime'. There is an algorithmic way of listing these. They can accordingly be written as Π_0, Π_1, Π_2, and so on. And we can consider an infinite table of *yeses* and *noes* registering whether or not successive natural numbers belong to the sets characterized by successive expressions on this list. Here is what the top left-hand corner of the table might look like.

	0	1	2	...
Π_0	yes	no	no	...
Π_1	no	yes	yes	...
Π_2	no	yes	no	...
.	.	.	.	...
.	.	.	.	...
.	.	.	.	...

Now consider the infinite sequence of *yeses* and *noes* that constitute the diagonal of this table, starting at the top left-hand corner.

	0	1	2	...
Π_0	yes	no	no	...
Π_1	no	yes	yes	...
Π_2	no	yes	no	...
.	.	.	.	...
.	.	.	.	...
.	.	.	.	...

And consider the infinite sequence of *yeses* and *noes* in which these are reversed, which in this example is:

no, no, yes,...

This sequence of *yeses* and *noes*, just like the sequence of *yeses* and *noes* on any row in the table, corresponds to a set of natural numbers. Call this set **D**. Now if there is an algorithm for deciding whether or not an arbitrary arithmetical statement is true, then there is an algorithm for deciding whether there is a *yes* or a *no* at an arbitrary point in the table, and therefore for deciding whether or not an arbitrary natural number belongs to **D**. But the language of arithmetic is sufficiently rich to ensure that, if there is an algorithm for deciding whether or not an arbitrary natural number belongs to **D**, then there is an expression somewhere on the list that characterizes **D**. This, however, is impossible. Π_0 fails to characterize **D** with respect to 0. Π_1 fails to characterize **D** with respect to 1. Π_2 fails to characterize **D** with respect to 2. And so on. So there cannot be an axiomatization that captures all and only arithmetical truths. *Q.E.D.*

> **Gödel's second theorem**: Given any axiomatization that captures only arithmetical truths, the truths that it captures, if they constitute a sufficiently strong set, cannot include a statement corresponding to a statement of the set's consistency.

Proof

Let A be an axiomatization that captures only arithmetical truths. Let T be the set of truths that A captures. Let c be an arithmetical statement corresponding to the statement that T is consistent. And suppose T is sufficiently strong.

Now there is an arithmetical statement u that corresponds to the statement that u (itself) does not belong to T. If this statement belonged to T, it would be false. So it cannot belong to T, and is accordingly true. But although T does not contain u, its sufficient strength guarantees that it does contain the (weaker) arithmetical truth $[c \rightarrow u]$ corresponding to the conditional statement that, if T is consistent, then u does not belong to T.

Suppose now that c belongs to T. Then, since $[c \rightarrow u]$ belongs to T, their joint consequence u must likewise belong to T. But, as we have seen, u does not belong to T. Therefore c cannot belong to T. *Q.E.D.*

References

Gödel's article was originally published as 'Über Formal Unentscheidbar Sätze der *Principia Mathematica* und Verwandter Systeme I', in *Monatshefte für Mathematik und Physik* 38 (1931). Its translation by Jean van Heijenoort, as 'On Formally Undecidable Propositions of *Principia Mathematica* and Related Systems I', appears in Jean van Heijenoort (ed.), *From Frege to Gödel: A Source Book in Mathematical Logic, 1879–1931* (Cambridge, Mass.: HUP, 1967).

Chapter 1: What is Gödel's theorem?

The quotation by Roger Penrose is from his *Shadows of the Mind: A Search for the Missing Science of Consciousness* (Oxford: OUP, 1994), p. 64.

The quotation by Rudy Rucker is from his *Infinity and the Mind: The Science and Philosophy of the Infinite* (Princeton: PUP, 2nd edn, 2005), p. 158.

The quotation by David Foster Wallace is from his 'Approaching Infinity', in *Boston Globe*, 12 December 2003.

Chapter 2: Axiomatization: its appeal and demands

Benedictus de Spinoza's *Ethics* is in his *Spinoza: Complete Works*, ed. Michael L. Morgan and trans. Samuel Shirley (Indianapolis, Ind.: Hackett Publishing, 2002). The quoted basic principle is axiom 2 of part II.

Isaac Newton's *Mathematical Principles of Natural Philosophy* is in
his *Isaac Newton's Mathematical Principles of Natural Philosophy
& His System of the World*, trans. Andrew Motte and Florian
Cajori (Berkeley: UCP, 1946).

Euclid's *Elements* is published as his *Elements*, trans. and ed.
T.L. Heath (New York: Dover, 1956).

Chapter 3: Historical background

G.W. Leibniz's proof that $2 + 2 = 4$ occurs in his *New Essays on
Human Understanding*, trans. and ed. Peter Remnant and
Jonathan Bennett (Cambridge: CUP, 1981), book IV, ch. VII, §10.

Gottlob Frege's *The Basic Laws of Arithmetic* is published as his *The
Basic Laws of Arithmetic: Exposition of the System*, trans. and ed.
Montgomery Furth (Berkeley: UCP, 1964).

A.N. Whitehead and Bertrand Russell's *Principia Mathematica*
is published as their *Principia Mathematica* (Cambridge:
CUP, 1927).

David Hilbert sets out his programme in 'On the Infinite', trans. Stefan
Bauer-Mengelberg, in Jean van Heijenoort (ed.), *From Frege to
Gödel: A Source Book in Mathematical Logic, 1879–1931*
(Cambridge, Mass.: HUP, 1967).

Chapter 6: A second proof of Gödel's theorem, and a proof of Gödel's second theorem

Giuseppe Peano presents his axiomatization of arithmetic in his 'The
Principles of Arithmetic, Presented by a New Method', trans. Jean
van Heijenoort in Jean van Heijenoort (ed.), *From Frege to Gödel:
A Source Book in Mathematical Logic, 1879–1931* (Cambridge,
Mass.: HUP, 1967).

Chapter 7: Hilbert's programme, the human mind, and computers

The quotation from Gödel's article is at p. 615; emphasis in original.

The Lucas–Penrose argument is advanced by J.R. Lucas in his 'Minds,
Machines and Gödel', in *Philosophy* 36 (1961).

The Lucas–Penrose argument is advanced by Roger Penrose in his *The Emperor's New Mind: Concerning Computers, Minds, and the Laws of Physics* (Oxford: OUP, 1989), ch. 4.

The quotation about mathematical intuition by Gödel is given in Hao Wang, *From Mathematics to Philosophy* (New York: Humanities Press, 1974), p. 324, from the unpublished text of Gödel's 1951 Josiah Willard Gibbs Lecture; emphasis in original.

Further reading

Budiansky, Stephen. *Journey to the Edge of Reason: The Life of Kurt Gödel* (Oxford: OUP, 2021). The most recent and most comprehensive biography of Gödel.

Franzén, Torkel. *Gödel's Theorem: An Incomplete Guide to its Use and Abuse* (London: Routledge, 2005). An extremely reliable guide to Gödel's theorem, which is especially good at debunking common associated myths.

Goldstein, Rebecca. *Incompleteness: The Proof and Paradox of Kurt Gödel* (New York: W.W. Norton & Co., 2005). A lively and accessible account of Gödel's life and work.

Hofstadter, Douglas R. *Gödel, Escher, Bach: An Eternal Golden Braid. A Metaphorical Fugue on Minds and Machines, in the Spirit of Lewis Carroll* (Harmondsworth: Penguin, 1979). A hugely entertaining account of Gödel's theorem and much else besides, with a particular emphasis on issues about artificial intelligence.

Nagel, Ernest and Newman, James R. *Gödel's Proof* (London: Routledge & Kegan Paul, 1959). Similar to my own book, but with a somewhat different approach and different emphases.

Smith, Peter. *An Introduction to Gödel's Theorems* (Cambridge: CUP, 2nd edn, 2013). An outstanding and very readable introduction to Gödel's theorem, including discussion of its wider significance.

Smullyan, Raymond M. *Gödel's Incompleteness Theorems* (Oxford: OUP, 1992). An excellent introduction to Gödel's theorem and associated results.

Index

For the benefit of digital users, indexed terms that span two pages (e.g., 52–53) may, on occasion, appear on only one of those pages.

Entries in **bold** indicate where technical terms are defined.

W

Wallace, David Foster 11–12
Whitehead, A.N. 37–42
  see also Principia Mathematica
Wiles, Andrew 8

Ω

ω-(in)consistency 75, 80, **82**,
  90–1

NUMBERS
A Very Short Introduction
Peter M. Higgins

Numbers are integral to our everyday lives and feature in everything we do. In this *Very Short Introduction* Peter M. Higgins, the renowned mathematics writer unravels the world of numbers; demonstrating its richness, and providing a comprehensive view of the idea of the number. Higgins paints a picture of the number world, considering how the modern number system matured over centuries. Explaining the various number types and showing how they behave, he introduces key concepts such as integers, fractions, real numbers, and imaginary numbers. By approaching the topic in a non-technical way and emphasising the basic principles and interactions of numbers with mathematics and science, Higgins also demonstrates the practical interactions and modern applications, such as encryption of confidential data on the internet.

www.oup.com/vsi

GERMAN PHILOSOPHY
A Very Short Introduction
Andrew Bowie

German Philosophy: A Very Short Introduction discusses the
idea that German philosophy forms one of the most revealing
responses to the problems of 'modernity'. The rise of the modern
natural sciences and the related decline of religion raises a
series of questions, which recur throughout German philosophy,
concerning the relationships between knowledge and faith,
reason and emotion, and scientific, ethical, and artistic ways
of seeing the world. There are also many significant philosophers
who are generally neglected in most existing English-language
treatments of German philosophy, which tend to concentrate
on the canonical figures. This *Very Short Introduction* will include
reference to these thinkers and suggests how they can be
used to question more familiar German philosophical thought.

www.oup.com/vsi